Interactive TV Technology and Markets

For a listing of recent titles in the *Artech House Digital Audio and Video Library,* turn to the back of this book.

Interactive TV Technology and Markets

Hari Om Srivastava

Artech House
Boston • London
www.artechhouse.com

Data

Om Srivastava.
       p. cm. — (Artech House digital audio and video library)
    Includes bibliographical references and index.
    ISBN 1-58053-321-3 (alk. paper)
    1. Interactive television.   2. Television broadcasting.
    I. Title. II. Series.
    TK6679.3 .S75 2002
    384.55'5—dc21   2001055248

British Library Cataloguing in Publication Data
Srivastava, Hari Om
    Interactive TV technology and markets. — (Artech House
    digital audio and video library)
    1. Interactive television   2. Interactive television—Standards
    3. Interactive television—Marketing
    I. Title
    384.5'5
    ISBN 1-58053-321-3

Cover design by Gary Ragaglia

© 2002 ARTECH HOUSE, INC.
685 Canton Street
Norwood, MA 02062

International Standard Book Number: 1-58053-321-3
Library of Congress Catalog Card Number: 2001055248

10 9 8 7 6 5 4 3 2 1

Dedicated to my brave daughters, Aradhana and Swati

Contents

Preface

Radical changes are taking place in the television industry. Passive linear entertainment viewing on television by millions of people around the globe is becoming a thing of the past. Television has now become an on-demand, participatory, nonlinear means of information and entertainment. In a very short time, Internet capabilities and interactivity will become integrated with television content. Viewers will be able to control the type of content they watch, its timing, and the outcome of a movie or drama.

In the future, content will essentially be enhanced. To viewers, this enhancement will appear as graphical or informational overlay. In addition to video and audio, enhanced content will include a variety of supporting information visuals, such as player information and statistics to support sports programs, and more detailed pictorial and text information, such as maps and biographies to supplement news coverage. This may also include interactive content that enhances a program's educational and entertainment value or provides viewers with the ability to order on-line goods and services featured in programming.

Today, interactive television (ITV) is interchangeably referred to as "Internet TV," "Internet for TV," "WebTV," and similar terms that invoke the Internet and the World Wide Web. Interactive TV combines the appeal and mass audience of traditional TV with the interactive nature of the Web. It could allow users to view tonight's news headlines while watching a favorite serial. Unlike traditional broadcast television, which transmits a one-way signal into people's homes, ITV lets viewers send messages back. This makes it possible for viewers to surf the Internet on the TV screen, send e-mail, or see specially prepared background information, such as a player's statistics in pop-up text boxes, that has been embedded into a TV program and can be seen by clicking on an icon with a handheld remote control. ITV makes it possible to check individual player and team statistics while watching a cricket show. The new technology also allows viewers to chat with TV talk show hosts or other viewers, play interactive games on their television set, and buy products via the TV with a click of the remote control. As analog TV signals fade out over the next 10 years, cable, satellite, and terrestrial TV network operators will use digital bandwidth to deliver more and more inter-active content to home-bound consumers, from purchasing pay-per-view programs and pizza to watching horse racing and shopping on impulse dur-ing commercials.

ITV is real. It is not a future dream. It is happening today. Several surveys, studies, and research firms promise great potential in ITV in the coming years. Cambridge, Massachusetts–based Forrester Research, Inc. predicts that television commerce (t-commerce) will be worth $28 billion by 2005 and account for about 16% of all on-line retail sales. The research group also predicts that ITV will be the key technology to bring e-mail and other on-line interactive services into homes that do not yet have Internet access.

This book has been written to illustrate interactive TV's roadmap. Chapters cover developments in broadcasting, an understanding of ITV technology, standards and standards bodies, a feel of the content creation for enhancement, the market for ITV, and the technology's future.

Chapter 1, on broadcasting, describes the various aspects of broadcast-ing, features of traditional broadcasting, compression techniques, digital broadcasting, flexibility of digital systems, tools and techniques for digital broadcasting, and the social aspects of broadcasting.

Chapter 2 describes the definition, feel, and history of ITV's develop-ment. It also discusses lessons learned from earlier trials, future promises, and the utility and offerings of ITV.

Chapter 3 deals with the technology and standards for ITV. Among the topics discussed are the various components of ITV; its technology, includ-ing conditional access (CA) and application programming interface (API); the use of "television blanking intervals" for sending data; and the return channel for getting data back from users. The chapter also provides details about worldwide standardization efforts and the organizations involved in this process.

Chapter 4 describes the content development techniques for ITV. Content refers to the media material available to viewers via the system. The distinguishing feature of this new media production versus the old is the inclusion of interactivity into the creative palette. This new aesthetic has given rise to the promotion of new media vocabularies in creating and interpreting programs and messages, games, news, services, and advertisements. The formation of this "new language," which did not form exclusively through the discourses of professionals, is as sensitive to the technology or platform as it is to the understanding and interpretative abilities of the viewers. The development of digital interactive technology for domestic and personal use is arguably the phenomenon of the decade. Adding enhanced content to TV programming and commercials opens up a wide range of new business models. Advertisers can supplement their linear video commercials with detailed or customized information and even enable viewers to order on the spot. At the same time, advertisers can leverage their existing Web sites by combining them with enhanced advertising, and programmers can develop enriched content that draws new viewers while retaining old ones. In Europe, ITV has taken off with satellite services offering interactivity. In the United States, there are 5 million subscribers to ITV services, and Forrester estimates that this number will increase to 65 million by 2005. Forrester further projects that, by the end of 2005, ITV services in the United States will generate $7 billion in subscriptions, $17 billion in marketing and advertising fees, and $23 billion in e-commerce. Chapter 5 discusses the potential for t-commerce and also the broad issues governing business models of this new media.

ITV's future is discussed in Chapter 6. Bandwidth limitations place restrictions on ITV's current capabilities. As digital TV becomes more common, and high-speed cable, satellite, and DSL technologies make their way into more homes, the interactive TV experience is bound to change. However, the most important parameter is human psychology and behavior. ITV promises new creative possibilities in content and commercials. But these will have to be based on human behavior, which is subject to cultural, ethnic, educational, and economic backgrounds. The rewards, for those who embrace the future and accept the challenge of learning what it takes to survive and prosper by new rules, are there for the taking. Chapter 7 presents several case studies.

At the end of each chapter, references have been provided. A glossary covers terms used in ITV development, from technology, design, audience and user research, business, human factors, advertising, and social science research. The glossary provides basic definitions and an expansion of acronyms.

The book contains seven end-of-chapter appendixes. Appendix 2.A provides the name of partners in ITV trials in 1994–1996. Appendix 4.A provides glimpses of ITV link attributes and their syntax. Appendix 4.B provides a list of major enhanced TV developers, producers, and consultants. Appendix 4.C provides an overview of WebTV-supported syntax. Appendix 5.A is on projected investments on ITV infrastructure. Appendix 5.B provides Web sites of international companies dealing with interactive media. Appendix 6.A is a time line of ITV and related technology developments. This book has been written with the objective of providing the reader—who may be either a user, developer, or decision maker—with an overview of all aspects of ITV.

Acknowledgments

Many people other than the author have contributed to this book. I would especially like to acknowledge the contribution of Mr. Lokesh Kumar in preparing a number of the figures for the book. I also want to acknowledge the contribution of Mr. Arch C. Luther for his support in bringing the book to its final form through his valuable comments and suggestions. I would further like to thank Ruth, Judi, Shubha, Anil, Meena, and many others who helped me prepare the manuscript. In addition, I am grateful to Sony, Nokia, Wink, and Bosch for granting permission to use their photos. I also wish to acknowledge the contributions of many authors, researchers, and survey organizations, whose material on the Web provided me with important information while writing this book.

1

Broadcasting

1.1 Introduction

Radio and television broadcasting are the oldest and most successful electronic mass media. Traditional broadcasting services are universally available throughout the world, generate the largest advertising revenues among electronic media, and occupy about a third of the average person's waking hours. Even so, broadcast technology is undergoing a phenomenal change. During the last 10 years, sound and television broadcasting have undergone a revolutionary change from established analog to digital technology. New services, including high definition television (HDTV), digital audio broadcasting (DAB), digital video broadcasting (DVB), direct-to-home (DTH), and new direct broadcast satellite (DBS), are augmenting, and in some cases competing with, existing AM and FM radio, VHF, and UHF television.

The advent of digital technology, with the possibility of nonlinear editing, virtual studios, disk-based storage, signal processing, multichannel transmission, multiple delivery systems, and television commerce, points to a fascinating era. A new service of broadcasting—cyberspace—has recently emerged and is being tried by a number of broadcast organizations. The service provides flexible interactivity, where the listener/viewer can obtain the audio, video, and animation on demand. Enhanced and interactive TV (eTV and ITV) are other emerging markets that allow the viewer to control the type and timing of viewing/listening. New projects are constantly being undertaken to devise a consumer receipt system that may be a lingua franca

1

of education, entertainment, commerce, research, and personal information. The set-top box may, if not completely operate as both a TV and computer, certainly perform certain computer functions. All of these digital TV services and features are discussed in this book.

Broadcasting continues to change every day as the information and media industries plunge through a dynamic and volatile era. Familiar market sectors are converging and breeding new sectors, including the Internet marketplace. The convergence of these media technologies is opening many opportunities for new revenue channels. Broadcasting is emerging into a system with many players and partners delivering products, with enormous possibilities for commerce via both radio and TV systems.

With this in mind, it is important to know how the technology in the field of broadcasting is developing and how it is going to open up new revenue streams, not only for broadcasters but also a host of other players who will be partners in the revolution ahead.

1.2 What Is Broadcasting?

Broadcasting service deals with point-to-multipoint communications. Broadcasting traditionally encompassed radio and television, which provided entertainment, information, and education to the masses. Traditional broadcast used to be a "one-to-many" system, where the audience had to be a passive listener or watcher. The concept of data broadcasting as a value-added service was then introduced. In recent times, broadcasters are considering multimedia broadcasting, which consists of delivering audio, video, data, and text through a host of distribution systems such as terrestrial broadcast networks, satellites, asynchronous transfer modes (ATM), synchronous digital hierarchy (SDH), fiber cables, and the Internet.

1.2.1 Communication System

As the technological development of broadcast systems is discussed, it will be seen that the basic communications system has remained the same with three distinct components: the transmitter, channel, and receiver (Figure 1.1). The components functions are described below.

1.2.1.1 Transmitter

The transmitter's prime function is to generate the RF carrier frequency to be transmitted. This carrier is generated either through the use of a crystal

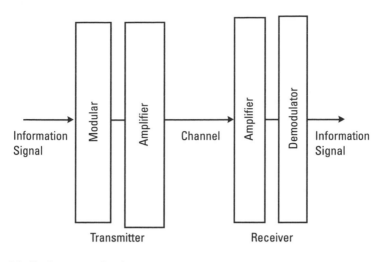

Figure 1.1 Basic communications system.

oscillator or the use of frequency synthesis (phase locked loops). In most modern transmitters, it is the phase lock loop that generates the carrier. This generated carrier is modulated, amplified, and then transmitted through the airwaves by use of an antenna network.

The purpose of modulation is to modulate the carrier with the desired information to be transmitted. In an amplitude-modulated system, the amplitude of the carrier is varied according to the modulating signal. For amplitude modulation, the modulation is accomplished in the final power amplification stages. For frequency modulated systems in which the carrier's frequency varies in accordance with the modulating signal, it is usually done at the second oscillator, if it is not synthesized. If the system is using a synthesized FM modulation, then phase modulation after the first oscillator is used.

Nonsynthesized transmitters (directly using the crystal as the source) need to multiply the signal as well as increase its amplitude. The two functions are performed using medium power amplifiers. The amplifiers are arranged in a chain, with each stage having more gain than the stage before. The first few stages are operated in the nonlinear region to generate harmonics. With FM synthesized transmitters, output of the second oscillator is amplified, and this oscillator runs on the correct transmitter frequency.

Once the signal is at the correct frequency, amplified to desired level, and modulated with information to be transmitted, it is applied to an antenna without losing any power. The signal is routed through a matching circuit to bring the impedance to the value necessary (typically 50 ohms for mobile radio). The signal is passed through the antenna switch to a harmonic filter and then to the antenna.

The digital signal arises from several sources, namely encoding ASCII characters, sampling and quantizing speech, or a video signal. The high-quality video input is taken via a signal-processing unit before being applied to the source encoder. The output data is subjected to forward error correction (FEC) channel coding. In the subsequent multiplexer, the audio, sync data, and control signals are added to the coded picture signal. The channel-coded data are modulated onto the IF carrier and packed into a channel bandwidth. A mixer up-converts to the RF carrier frequency before the signals are applied to the terrestrial transmitter and antenna. We shall discuss the process of digital modulation and encoding in Section 1.6.

1.2.1.2 Channel

The medium, which transports the modulated signal to the receiver, acts as the channel for medium wave (MW), shortwave (SW), and frequency modulated (FM) radio broadcasts. Modern broadcast, such as cable TV or the Internet, also uses a number of other mediums as channels.

1.2.1.3 Receiver

The subsystem takes in the transmitted signal from the channel and processes it to retrieve the information. The receiver must be able to discriminate the signal from other signals that may be using the same channel, amplify the signal for processing, and remove the carrier to retrieve the information. The desired signal is selected against unwanted signals in the channel by a method known as tuning. Similarly, retrieving the information signal is accomplished by a technique known as demodulation.

Heterodyning is a process where two frequencies are mixed together to produce a beat frequency, namely the difference between the two. Amplitude modulation is a heterodyne process, where the information signal is mixed with the carrier to produce two side-bands. The side-bands occur at precisely the sum and difference frequencies of the carrier and modulating signal. These are beat frequencies. Superheterodyning is simply reducing the incoming signal frequency by mixing. In a radio application, the AM or FM signal centered on the carrier frequency is reduced to an intermediate frequency (IF). For practical purposes, the superheterodyne receiver always produces the same IF. To accomplish this, the frequency being mixed into the signal is continuously varied so as to keep the difference the same. The IF is demodulated, amplified, and fed to the speaker or display unit for reproduction of sound and images. We will discuss the process of receiving digital reception systems in Section 1.6.

1.2.2 History of Broadcasting

1.2.2.1 Radio

The era of radio broadcasting began in 1901 when Guglielmo Marconi achieved dramatic success in transmitting radio signals across the Atlantic Ocean. By 1921, medium wave (531–1602 kHz, extended to 1700 kHz in North America) radio broadcasting had become a reality. Shortwave (3–30 MHz) broadcasting was further demonstrated by Marconi in 1922.

The powerful impact radio can have on listeners' imaginations was amply shown on an evening in October 1938, when a direct broadcast of the one-hour play *War of the World*—produced by Orson Wells from Mercury Theatre—stunned the world [1]. The channel congestion on medium wave led to the use of VHF for broadcasting. The first stereo broadcasting on FM for high-quality audio was introduced in 1962. The era of digital audio broadcasting started in 1995 with the regular transmission of digital broadcasting in Europe. The chronology of events for sound recording and a history of broadcasting appear in Table 1.1.

Radio Broadcasting Service

The broadcasting service is used to serve both international and domestic audiences. International audiences are served by *external services,* which are mostly shortwave and medium wave radio transmitters. Domestic audiences are served by *home-service* over-the-air radio systems. This service is also the major source of local news, sports, public affairs, and many other programs.

External Service External service by its very nature requires the generation of signals that are intended to be transmitted across international borders. Consequently, transmission of these signals is subject to the International Telecommunication Union's (ITU) radio regulations. For decades, governments have made increasing use of the electromagnetic spectrum to conduct public diplomacy by broadcasting speech and music throughout the world. The external services are based on shortwave transmission, which reaches a wide geographical area using ionospheric propagation. However, with the advent of satellite broadcasting, digital satellite radio systems are being used to cover large areas. These technologies will be discussed later.

Table 1.1
Brief History of Sound Recording and Broadcasting Technology

1864	James Maxwell's electromagnetic wave theory for radio wave propagation.
1877	First description of recording sound onto a cylinder or disc described by Charles Cros in France and Thomas Alva Edison in the United States.
1878	Thomas Edison patents the recording of sound onto discs and cylinders.
1887	Heinrich Hertz transmits and receives radio waves over short distances.
1888	Emile Berliner shows first example of a working "phonograph" playback device.
1888	Basics of magnetic recording put forth by Oberlin Smith.
1889	Danish inventor Valdemar Poulsen patents the first magnetic recorder.
1895	Development of first wireless telegraph system by Guglielmo Marconi.
1905	First electron tube developed by Sir Ambrose Fleming.
1906	First wireless communication of human speech.
1919	KDKA in Pittsburgh, Pa., is licensed as the first broadcast radio station.
1925	First electronic recordings made with the use of a microphone.
1931	First experimental stereo recordings made by Bell Telephone Laboratories.
1933	Theory of frequency modulation (FM) for radio broadcasts by Edwin Armstrong.
1948	Introduction of long play (LP) record by CBS.
1948	First transistor introduced.
1954	Introduction of stereo tapes to the public.
1962	First stereo FM radio broadcasts.
1975–1978	Early digital recording made.
1980	Sony introduces the "Walkman."
1983	First CD player made available by Sony and Philips.
1995	Digital broadcasting starts.

Home Service   Domestic audiences are served by AM and FM broadcast stations employing analog radio transmissions designed for direct reception by home receivers. AM broadcast stations operate on a channel in the 531–1602 kHz AM broadcast band. This band consists of 120 carrier frequencies beginning at 531 kHz and progressing in 9 kHz steps to 1602 kHz. The modulation of the radio carrier wave is amplitude modulation; hence, the AM reference. The operating power ranges from 1 to 1000 kilowatts. Propagation in the AM broadcast band involves both the ground wave and sky wave modes, which serve local and distant audiences, respectively. A disadvantage in AM broadcasting is its limited audio fidelity, relative to FM.

FM broadcast stations are authorized for operation on 80 allocated channels, each 180-kHz wide, extending consecutively from 100 (88 in the United States) MHz to 107.9 MHz, with a standard deviation of 75 kHz. The effective radiated power ranges from a few watts to 50 kilowatts. Better audio fidelity is a distinct advantage of FM radio over AM radio broadcasting; however, FM radio does not normally have the extensive service coverage areas that AM radio broadcasting enjoys.

1.2.2.2 Television

Almost every home has a television set. They are maligned, lauded, criticized, and praised, but they are watched. Television is regarded by some as mankind's greatest invention, possibly more life sustaining than fire and certainly more entertaining than the wheel. In its early days, it was regarded by many as the anti-Christ; and that attitude began with the transmission of the very first television picture in 1926.

The television era started with video broadcasting in 1936. Television started early on in the history of radio but did not really take off until the 1950s. Color was added to TVs in 1950. The National Television Systems Committee (NTSC) color standard was adopted in the United States in 1953, and Europe followed with the Phase Alternating Line (PAL) standard in 1966. When color became common in the early 1960s, the stage was set for explosive growth. Since then, there has been no looking back. Television development continued with the introduction of UHF transmitters in 1952 to address the problem of spectrum limitation in VHF band. The launch of Sputnik in 1957 started the space race. With the formation of Intelsat by the International Satellite Organization, a new era of global connectivity started. In 1975, the delivery of video programs to cable head ends via satellite started. The demand of video entertainment by viewers not served by cable gave rise to the DTH delivery system, beginning in 1984, which allows users to receive signals directly from satellites via a small 30-cm dish.

In 1993, experiments on digital television in the terrestrial 8/7 MHz channel started. Digital satellite system (DSS) was introduced in 1995. A brief history of video and television technology is given in Table 1.2.

Today, television is the world's most powerful form of communication. Every day, it reaches out to millions of people to entertain and inform them with "real-life" images of the world around them.

How Does Television Work? Television pictures come from electronic signals like radio waves. The color television camera begins the process of creating a picture on the television screen (Figure 1.2). This camera focuses images on

Table 1.2
Brief History of Video and Television Technology

1897	Development of the cathode ray tube by Ferdin and Braun.
1907	Use of cathode ray tube to produce television images.
1923	Patent for the iconoscope, the forerunner of the modern television pickup tubes.
Early 1930s	RCA conducts black and white broadcasting experiments.
1936	First television broadcast made available in London.
1938	Initial proposal for color television broadcast made by George Valensi.
1949	System developed to transmit chrominance and luminance signals in a single channel.
1954	NTSC standard for color television broadcast introduced in the United States.
1966	PAL standard introduced in Europe.
1975	Sony markets the first Betamax VCR for home viewing and recording of video.
1976	JVC introduces the VHS format to the VCR arena.
1976	Dolby Laboratories introduces Dolby stereo for movies.
1978	Philips markets the first video laser disc player.
1984	The first hi-fi VCR is introduced.
1985	The broadcast of stereo television.
1994	Standard agreed on for HDTV transmission.

television pickup devices, which convert light energy into electrical energy. The color television camera has three pickup devices—one for each of the primary colors, red, green, and blue. Unlike movie pictures, which are whole-screen images, television pictures are made up of horizontal lines. Each line is transmitted one at a time in a process called scanning. NTSC is a 525-line system. PAL uses 625 lines.

SECAM is a third standard. SECAM shares with PAL/625, the higher number of scan lines than NTSC/525. Instituted in 1967 by France, SECAM, short for "Sequential Couleur à Memoire," has an odd history. Sensing that they could develop their own market for television sets, the French resisted joining the rest of Europe by not adopting PAL and instead developing SECAM. However, it turns out that SECAM is so difficult to edit that it is often converted to PAL for editing. As for the French market for manufacturing TV sets, Asian-built tristandard TV sets have taken a large part of their home market. However, each standard has a few advantages and disadvantages. So, to get the moving picture in the PAL system, the color

Figure 1.2 Television camera. (*Source:* Sony. Printed with permission.)

television camera has to scan 625 lines in every image, 25 times every second. It takes the camera 64 millionths of a second to scan from side to side of one line in a single television picture. The television camera encodes the three separate color signals into one signal, which is sent from the camera either to a video recorder or direct to the station's transmitter. The transmitter sends out a broadcast signal to be picked up by television antenna. The TV set then converts the electrical impulse of that signal to a color image just like that picked up by the television camera.

The human eye retains an image for a fraction of a second after it views the image. This property, called persistence of vision, is essential to all visual display technologies. The basic idea is quite simple—single-still frames are presented at a high enough rate so that persistence of vision integrates these still frames into motion. Motion pictures originally set the frame rate at 16 frames per second. This was rapidly found to be unacceptable, and the frame rate was increased to 24 frames per second. In Europe, this was changed to 25 frames per second, as the European power line frequency is 50 Hz. When NTSC television standards were introduced, the frame rate was set at 30 Hz (one-half the 60-Hz line frequency). Then, the rate was moved to 29.97 Hz to maintain 4.5 MHz between the visual and audio carriers.

We will now discuss the process known as "interlacing" to avoid the "flicker." To explain this, we will use the PAL system. In early television sets, the process of writing 625 lines for each frame created noticeable flickering. To minimize the flicker, a system of "interlaced scanning" was developed. The interlace system divides each frame into two separate fields, each with half of the picture information, for a total of 625 lines of picture information. The first field contains all odd-numbered lines (i.e., No. 1, No. 3, and so on). The second field contains the even-numbered lines (i.e., No. 2, No. 4, and so on). After field one is scanned for all the odd-numbered lines, a vertical synchronization pulse returns the electron beam to the top center of the picture tube and then scans all the even-numbered lines. Each of the 25 frames of a video picture includes these two interlaced fields, so the actual scanning rate is 50 fields per second. If the scanning were not interlaced, the entire 625 lines would illuminate the screen 25 times per second. The eye starts to blend flashing pictures when the pictures scan at a rate of about 45 per second. This gives the viewer the impression of a continuously lighted image. If not interlaced, the resulting 25-times-per-second scan rate would flicker and be objectionable to the eye.

Television Signal Transmission   No matter how the signal is sent from one location to the other, be it radio waves, coaxial cable or fiber optics, the first thing to consider is the method used to convert images to electronic signals and back again. This process is the heart of television.

The first part of the process involves getting the image into an electronic form. The picture is broken up into small pieces by a scanning process and sent one line at a time. In early cameras, various forms of special tubes were used for this process. The tubes had a special layer that was sensitive to light. As an electron beam scanned this layer, the image that was focused on the tube was converted to an electronic voltage, which corresponded to the brightness of the image. Modern cameras use solid state devices that do essentially the same thing. Color images are sent in the same way, except that there must be a signal for each of the primary colors used.

TV stations operate on 6-MHz wide channels in the VHF and UHF frequency bands. The spectrum occupied by television broadcast comprises 72 MHz in the VHF band and 336 MHz in the UHF band.

Regular television stations use a form of AM modulation known as vestigial sideband (VSB) to send the video via radio waves. Similar to AM, VSB filters out duplicate parts of the modulated signal in order to reduce the amount of channel space, or bandwidth, necessary to send the signal. The signal is sent through terrestrial transmission systems using VHF or UHF antenna or satellite systems. Traditional satellite TV uses microwave frequencies that require large dishes for best performance. These systems use FM to send the video, which reduces interference between the various satellite transponders (satellite circuitry that relays the TV signal). The new DSS satellite TV units operate at a higher frequency than traditional units, which allows the dishes to be smaller. In addition, the video signals are converted to digital data, which is sent over the satellite link. At the receiver, the digital signals are converted back to video, which is sent on to the TV set. DSS signals are generally of higher quality than conventional satellite TV. We will discuss the DSS system in Section 1.6.7.

1.3   Broadcasting and Telecommunications

Broadcasting is traditionally defined as a mass communication system for addressing many audiences from one transmitter. A centrally-located transmitter is able to deliver messages to masses of people living in its target area. Broadcasting is normally an over-the air transmission system. The audience is a passive listener and watcher. Apart from broadcasting, telecommunications is the system for addressing one-to-one. The telephone is an example of a telecommunications system. The traditional concept is the use of a plain old telephone system (POTS) with twisted-pair cable reaching individual homes and using a telephone exchange system to physically establish a connection between two parties.

However, with the advent of digital compression systems, a number of protocols, namely the Internet Protocol (IP) and a variety of delivery systems, such as cable, satellites, and the Internet, the distinction between broadcast-type services and telecommunications services are blurring. All types of operators are turning into service providers. Different types of services, such as video, audio telephony, and text can be transmitted on a mix of delivery channels. Broadcasters are devising different ways to provide interactivity. The concept of a "me channel" is evolving, where the audience can have full control over viewing and content. A number of experiments are being done to provide a "return link" within the broadcast band or by using the telephone or the Internet. Broadcasters are trying to provide services such as e-mail, paging, or voice over IP (VOIP), which were previously the domain of telecommunications operators. On the other hand, telecommunications operators are providing services such as video on demand (VOD), as well as other broadcast services. The era of convergence is coming extremely fast.

1.4 Digital Television and Compression Technique

PAL, SECAM, and NTSC were adopted for composite color signals in the 1960s to maintain a compatible reception by millions of black-and-white TV sets. To keep the same channel allocation and save the frequency spectrum, the same black-and-white channel bandwidths were used; approximately 4.2 MHz for NTSC and 5.5 MHz for PAL/SECAM. Such a low bandwidth was found adequate to transmit to TV receivers, albeit with drawbacks like "cross color" in specific conditions [2]. However, these were not sufficient for processing in the studios. Consider, for example, chroma keys. Chroma key is an electronic process that allows two camera signals to be combined, using one as the background and the other as the foreground. A typical application is to show a weathercaster (foreground) standing in front of a computer-generated weather map (background). This is accomplished by having the weathercaster stand in front of a blue background; the chroma keyer senses the blue color in the foreground camera signal and inserts the background image from the computer into all blue areas of the foreground picture. Chroma key creates keys on just one color channel. Broadcast cameras use three independent sensors, one for each color, red, green and blue (Section 1.2.2.2). Most cameras can output these RGB signals separately from the composite video signal. The original chroma key was created by feeding the blue channel of a camera into a keyer. Manufacturers soon created dedicated chroma keyers that could accept all three colors, plus the background and foreground composite signal. This made it possible to select any color for the key and fine-tune the color selection.

A composite video signal is one in which the luminance and chrominance are combined with synchronization signals using an encoding standard (e.g., PAL, NTSC, SECAM). A component video signal is one in which the luminance and chrominance are sent as separate components, either as RBG or as luminance and color difference signals, (e.g., Y, Cb, and Cr). The component standards commonly known as "BetaCam" and the "M" formats evolved in 1981. This analog component standard had the disadvantage in post-production because it needed three cables for every signal instead of one for the composite. For these reasons, it was necessary to use a digital television standard that would make it possible to preserve the original quality whenever the processing became too complex. The advantages of digital technology over analog for TV are given in Table 1.3.

1.4.1 ITU-R 601 Component Digital

The process of capturing images generates analog signals. Converting them into digital bits involves two critical steps: sampling and quantization. As the term implies, sampling is the periodic measurement of analog values to produce image samples. Quantization is the process of converting the sampled voltage levels into digital data. The basic scheme is to predict motion from frame to frame in the temporal direction and then to use discrete cosine transforms (DCT) to organize the redundancy in the spatial directions. The DCTs are done on 8×8 blocks, and the motion prediction is done in the luminance (Y) channel on 16×16 blocks. In other words, given the 16×16 block in the current frame that is being coded, a close match to that block in a previous or future frame is searched (there are backward prediction modes where later frames are sent first to allow interpolating between frames). The DCT coefficients (of either the actual data or the difference between this block and the close match) are "quantized," which means that they are divided by some value to drop bits off the bottom end. Many of the coefficients then end up being zero. The quantization can change for every "macroblock" (a macroblock is 16×16 of Y and the corresponding 8×8s in both U and V). The results of all this, which includes

Table 1.3
Advantages of Digital over Analog Technology for TV

Characteristics	Remarks
Picture quality	Does not deteriorate after multiple reproductions
Sound quality	Does not deteriorate after multiple copyings and reproductions
Multichannel	It is possible to transmit multiple channels in the same bandwidth
Reception quality	Reliable and robust. Stays good as signal strength lowers, until a threshold is reached where the output signal abruptly disappears.
Multiprogram	Provides possibility of mixing of audio, text, data, and multimedia on the same channel.
Interactivity	Better suited for interactivity
Terminal equipment	PC or TV

the DCT coefficients, the motion vectors, and the quantization parameters, is Huffman coded[1] using fixed tables. The DCT coefficients have a special Huffman table that is two-dimensional in that one code specifies a run-length of zeros and the nonzero value that ended the run.

Nyquist frequency is the lowest sampling frequency that can be used for analog-to-digital conversion of a signal without resulting in significant aliasing. Normally, this frequency is twice the rate of the highest frequency contained in the signal being sampled. The Nyquist limit determines the bandwidth of sampling frequency.

ITU-R 601 was defined as an international standard for component coding of TV signals [3]. It specified orthogonal sampling at 13.5 MHz for luminance (Y) and 6.75 MHz for the two-color difference signal Cb and Cr. The sampling specification is based on multiples of a 3.375-MHz sampling rate, a number carefully chosen because of its relationship to both NTSC and PAL. Two levels of quantization are permitted for the luminance component: 8- and 10-bit. The color difference components are quantized at 8-bit. The sampling frequency for the luminance (Y) component of the signal is 13.5 MHz. Thus, the upper limit on the spatial frequencies that can be represented will be slightly less than 6.75 MHz. This is sufficient for the 6.0-MHz luminance bandpass of PAL and represents significant horizontal oversampling relative to the 4.2-MHz luminance bandpass for NTSC.

The sampling frequency for the color difference components, R-Y and B-Y, is 6.75 MHz. Thus, the upper limit on the spatial frequencies that can be represented is slightly less than 3.375 MHz. This is significantly higher than the 1.5-MHz bandpass for the color components in both PAL and NTSC. This additional color resolution is quite useful in the generation of color mattes, providing yet another reason to use component rather than composite video signals in the video composition process. Color matte is a color signal that may be adjusted for chrominance, hue, and luminance. Matte is used to fill areas of keys and borders. The choice of sampling frequencies gives 720 samples per active line for luminance Y and 360 samples for color differences. It also includes space for representing analog blanking.

To ease the handling of all the frequency ratios described above, a notation system has been adopted to describe the relative sampling frequencies for the luminance and color difference components of the digital video signal; this notation is referenced as the 3.375-MHz basis for the ITU-R 601 standard. The luminance sampling frequency is four times the base

---

1. Huffman is a method for encoding messages.

frequency; the color difference sampling frequency is two times the base frequency. Thus, this sampling relationship is know as 4:2:2, representing luminance and each color difference component. These sampling parameters are used in the D-1, D-5, digital BetaCam, digital-S, and DVCPRO50 tape formats.

The notation 4:4:4 indicates that each of the signal components is sampled at 13.5 MHz. This corresponds to the signals produced by the individual image sensors, thus the 4:4:4 notation is often used for either YUV or RGB components.

The notation 4:1:1 indicates that the color difference components have one-quarter the resolution of the luminance signal; this correlates well with luminance and color resolution delivered by NTSC and PAL. 4:1:1 sampling is used in the consumer DV format and the 25-Mbps versions of DVCPRO and DVCAM.

1.4.2   MPEG Standardization Development

Digital video and audio require huge space for storage and transmission. It is simply impractical to store and transmit the digital signal without some compression or encoding. Digital video and audio coding technology has a long research history, and international standards have also been established. The Moving Picture Experts Group (MPEG) is a joint technical committee of the International Organization for Standardization (ISO) and the International Electrotechnical Commission (IEC). MPEG-1, MPEG-2, and MPEG-4 standards have already been developed. Standardization of the MPEG-7 system will follow the above-mentioned systems, and it will regulate content description data formats that support video and audio content retrieval.

1.4.2.1   MPEG-1

MPEG-1 is the first video-coding standard completed by MPEG. It is a coding format for storage media such as CD-ROMs. It compresses video data with half the resolution of standard definition television (SDTV) both vertically and horizontally (SDTV is discussed further in Section 1.6.11). MPEG-1 has parameters of 30 SIF (source input format) pictures (352 pixels by 240 lines per second) down to approximately 1.5 Mbps. The number of quantization levels determines the accuracy with which a sample can be represented. More quantization steps provide greater accuracy and the ability to reproduce the small levels of difference between samples in smooth gradients.

Using a technique called "motion compensation prediction," MPEG-1 first identifies frames with different image patterns from consecutive frames.

There are two methods of motion compensation: one estimates from the previous frame and the other estimates from the previous and proceeding frames (bidirectional prediction). In addition, MPEG-1 classifies this selected information using a DCT technique based on the relevance to human visual perception (see Section 1.4.1). By omitting extraneous information using human visual peculiarity, drastic data compression is accomplished by means of MPEG-1. MPEG-1 picture quality is inadequate for regular broadcasting due to the significant resolution reduction. A few key features are given in Table 1.4.

1.4.2.2 MPEG-2

MPEG-2 has now become the world standard for broadcast quality video. MPEG-2 was designed with the aim of expanding the general-purpose coding scheme, making it applicable to various other media fields, including broadcasting. Its standardization was advanced in cooperation with other international standardization organizations, such as the ITU-R and the ITU-T. The MPEG-2 system manages a wide range of video input images, from SIF to HDTV, which is further explained in Section 1.6.11. The data compression technique is based on the MPEG-1 format, and it also includes a system that efficiently codes interlaced images (which comprise a normal television signal). Furthermore, it introduced a hierarchical (scalable) coding function, enabling a low-quality partial reproduction of coded information to be completely reproduced.

Since MPEG-2 covers a wide range of applications, some of its functions are unnecessary, depending on how it is used. Therefore, multiple subset standards are defined within the system and used by the application of two indexes: one is a profile index that indicates the degree of functionality,

Table 1.4
MPEG-1

Development	January 1988–November 1992
Application	CD-ROM, Video Karaoke
Input	SIF video (horizontal 352 pixels, vertical 240 pixels)
Characteristics	Editing/random access feasible MPEG-1 standard designed to store video data on CD-ROM; later became a foundation for MPEG-2 and newer standards

and the other is a level index that establishes object image resolution. For example, the above-mentioned bidirectional prediction cannot be accomplished in a simple profile but only in a main profile. A main level is able to handle video inputs up to the point of SDTV, and a high level up to HDTV images. The MPEG-2 standard has been adopted for use in digital broadcasting worldwide. A few key features are given in Table 1.5. MPEG-2 levels and profiles are described in Table 1.6.

1.4.2.3 MPEG-4

MPEG-4 is a standard for multimedia communications. In addition to conventional video and audio coding capabilities, the MPEG-4 standard also features new functionalities, such as object-based video (Figure 1.3), scalable coding, error resilience, and so on, which make the multimedia information more attractive, interesting, and robust.

The primary goal of MPEG-4 is to provide a highly efficient video-coding scheme for transmission and to represent more attractive multimedia data with the capability of content-based interactivity, universal access, and error robustness. MPEG-4, whose formal ISO/IEC designation is ISO/IEC 14496, was finalized in October 1998 and became an international standard in the first months of 1999. The fully backward compatible extensions under the title of MPEG-4 Version 2 were frozen at the end of 1999. It acquired formal international standard status in early 2000.

Table 1.5
MPEG-2

Development	July 1990–November 1994
Application	Digital broadcasting, DVD
Input	SIF-SDTV-HDTV video
Characteristics	General-purpose coding with varied resolution and functions. Addition of functions for studio production developed in January 1996. Addition of stereo video coding function developed in October 1996. MPEG-2 is a general-purpose coding standard that absorbed data hierarchical functions in addition to managing SDTV and HDTV video data. Digital broadcasting throughout the world has adopted the MPEG-2 system.

Table 1.6
MPEG-2 Levels and Profiles

	Simple	Main	SNR Scalable	Spatial Scalable	High	Multiview	4.2.2
High level	—	X	—	—	X	—	—
High-1440 level	—	X	—	X	X	—	—
Main level	X	X	X	—	X	X	X
Low level	—	X	X	—	—	—	—

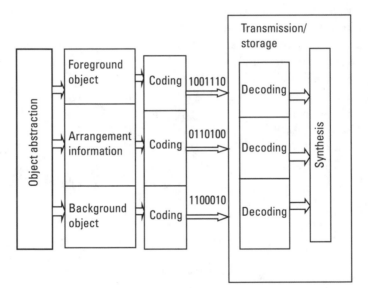

Figure 1.3 Object-based coding system.

MPEG-4 supports video encoding from 5 Kbps up to about 2 Mbps and can go as low as a few kilobits per second for graphical objects. For audio streaming, MPEG-4 incorporates scalable audio profiles to cope with variable bandwidth and sound quality, allowing quality to increase seamlessly as bandwidth becomes available and to decrease gracefully when bandwidth is reduced.

The standard is not a direct improvement of MPEG-1 or MPEG-2 coding standards, but rather defines a different way of integrating and interacting with scene content. This standard is centered around four ideas:

- Modular integration of images, text, audio, and video objects;
- Transparent delivery of content from a wide variety of sources;
- User interactivity;
- Identification and protection of intellectual property rights (IPR).

Presentation of Content

Media Objects   In MPEG-4, the elements in a scene are described using a tree-like structure. At the root of the tree, the composite scene exists. This composite scene is then broken down into individual objects (such as a person or a movie clip) and then into atomic structures called media objects (such as the person's video and audio channels) that make up the leaves of the tree. Each media object carries its own object description (OD). The media objects can be classified into four categories:

- *Natural Video Objects.* These objects include recorded or live video feeds and still images (for example, GIFs, JPEGs). MPEG-4, however, adds a new functionality of "shape coding"—the ability to encode or decode a video object of arbitrary shape as opposed to traditional rectangular encoding and decoding. This allows separation of moving video objects (or sprites) from a still-picture background.
- *Synthetic Video Objects.* These include 2D graphics and 3D graphics, with added focus on two aspects:
  - A parametric description of a synthetic description of the human face and body, as well as their animation.
  - Static and dynamic mesh coding with texture mapping.
- *Natural Audio Objects.* These include recorded or live audio feeds. MPEG-4 standardizes natural audio coding at bit rates ranging from less than 2 Kbps to over 64 Kbps.
- *Synthetic Audio Objects.* These include text-to-speech coding and synthesized audio. In MPEG-4 audio, one can change the pitch or the speed of the sound without either affecting the other.

Scene Description   In addition to providing support for coded individual objects, MPEG-4 also provides facilities to compose a given set of objects into a scene (Figure 1.4). If an OD of a primitive media object describes the decoding properties of the object, an extra stream of information independent of the OD describes where the object is placed on the scene. This separa-

Figure 1.4 Composing objects in a scene.

tion of the media object from its relative place in the scene allows a user to move an object around without affecting the rest of the scene.

Applications   In terms of low-bit-rate applications, there are home video cameras, Internet video distribution, and next-generation portable video-phones, some of which are already in practical use. A table of product examples is given in Table 1.7. There is also a vision of using MPEG-4 for mobile-media digital satellite broadcasting.

Details about MPEG-4 are given in Table. 1.8.

1.4.2.4   MPEG-7

MPEG-7, officially called "multimedia content description interface," is the fourth MPEG standard, after MPEG-1, MPEG-2, and MPEG-4. While the

Table 1.7
MPEG-4 Product Examples

Field	Products
Communication	Video telephony, video conferencing
Education	Remote classroom
Healthcare	Remote expertise (such as medical assistance)
Multimedia messaging	E-mail, video answering machines
Entertainment	VOD, games, multimedia presentations, enhanced electronic program guide (EPG)

Table 1.8

MPEG-4

Development	November 1994–Present
Applications	Portable videophone, Internet video, studio production
Input	Quarter Common Interchange Format (QCIF—horizontal 176 pixels, vertical 144 pixels) to HDTV video
Characteristics	Computer graphics (CG) animation coding, coding of individual component elements. Version 1 (fundamental part) standardized in May 1999. Version 2 (function addition) standardized in March 2000. Version 3 (studio production standard) is under examination.

earlier three compression coding standards were developed to code actual video and audio data, the MPEG-7 standard covers the content description data that is attached to multimedia data in various other formats, including those of MPEG-1, MPEG-2, and MPEG-4. Content description data, sometimes referred to as metadata or an index, is a description of what is included within the data. The details of the MPEG-7 standard are provided in Table 1.9.

Standardization Items   One of the standardization items is the "descriptor." The descriptor is a function that defines and forms the content description

Table 1.9

MPEG-7

Development	1996–Present
Applications	Program archives Program material database Home server Music database Electronic catalog
Characteristics	Standardization of contents description data. Scheduled to be approved in September 2001 as an international ISO standard. MPEG-7 is the fourth MPEG standard. The aforementioned MPEG-1, MPEG-2, and MPEG-4 were compression coding standards for actual video and audio data. MPEG-7 is a standard that addresses content description data that is attached to actual video and audio data.

data. The descriptor standardizes the meaning of each item and the data format that should be described. To have the above-mentioned descriptor adopted as a standard, it is essential to verify its efficient performance using actual video contents. The MPEG is pursuing the establishment of the standard while simultaneously conducting verification tests.

1.4.2.5 MPEG-21

The MPEG is about to start working on a new standard, MPEG-21, in expectation of the formation of an audio-video contents distribution market. Although the standard's parameters are not yet concrete, discussion is under way regarding copyright protection technology for content circulation [4].

1.4.2.6 MHEG

The MHEG, for hypermedia information coding expert group, is an ISO standard designed to meet the requirements of multimedia applications and services, running on heterogeneous workstations, that interchange information in real time. Examples of these applications include computer-supported multimedia cooperative work, multimedia message systems, audio-visual telematic systems for training and education, simulation and games, VOD services, ITV guides, and others. MHEG seeks to fulfill these interchange requirements by defining the representation and encoding of final-form multimedia and hypermedia information objects [5].

1.5 Mode of Delivery: Terrestrial, Cable, and Satellite

Transmission systems in the broadcast scenario deal with the distribution of signals via terrestrial, satellite, or cable. Until early 1990, the main delivery of broadcast signals was through terrestrial analog transmitters. Satellites were used to link terrestrial transmitters. The development was mostly toward increasing efficiency and signal quality compared with the physical links. In the 1980s, the cable system gained prominence, making it possible to deliver multichannel TVs to homes with minimal cost. The recent trend is toward digital broadcasting using terrestrial modes or direct satellite broadcasting for the DTH service.

People love television. In addition to providing choice, it has revolutionized the world and the way we communicate. The next generation of digital television will continue to significantly influence society. Broadcasting is a late starter in digitization, following behind other systems such as data communications, voice telecommunications, and mobile communications. In these areas, digital technologies have been steadily applied during the last 20 years or so, massively improving the reliability, quality, and capacity of services.

Television has lagged behind in the switch to digital technology for a number of technical, economic, and political reasons. However, now that digitization is finally being applied to television, already the most widely used communications medium on the planet, the impact will likely be revolutionary. For example, the capacity increases in broadcast output are dramatic with the current developments in compression technologies. The current spectrum devoted to TV broadcasts can support some 500 TV channels. Terrestrial DTV broadcasting is a wireless service, which occupies a big chunk of the electromagnetic spectrum. Yet with all this bandwidth, terrestrial broadcasting cannot hope to compete with cable and DBS, which have the capability to deliver more than 500 channels (at an average bit rate of 3–5 Mbps per channel). DTH via satellite can provide all these channels to viewers. When combined with the commercial and social power of the medium—there are currently 1 billion TV sets in use around the globe, which is more than telephone main lines—this represents a communications "big bang" without precedent.

The commonly accepted wisdom is that to win in the digital future, broadcasters have to offer more and better content. However, in order to win consumers, accessing and controlling that content has to be made easy, even fun. Watching entertainment on screen, especially with the many choices available in a digital environment, can become an unpleasant ordeal if too many options are presented. The saving grace is a content interface that makes the retrieval process efficient and intuitive. This happens by building on familiar models; in effect, adjusting technology to humans, instead of asking humans to make radical changes. Within the familiar context of TV, people will happily, gradually evolve how they consume, rather than make abrupt changes. That evolution has been happening since the medium was introduced. Consumers started out by watching TV closely, almost as if they were glued to the set and mesmerized by its novelty. Now faced with an overwhelming array of hundreds, if not thousands, of choices, consumers have to be in charge to be comfortable.

Digital broadcast technology is being used for the following:

- Digital TV;

- Digital radio;
- Data broadcasting.

In a digital domain, the signal is in the form of bits, which have only two values, zero and one. For digital transmission systems, it is immaterial whether these bits belong to audio, videotext, or graphic signals. As such, they are all delivered in the same way as digital data and distinct from an analog signal. They can also be broadcast via satellite, cable, and terrestrial means. However, the method of receiving this signal will vary.

1.6 Digital Television Standards

After numerous trials and experiments, digital television and audio have now become a reality, with a number of countries already broadcasting in digital mode. There are currently 11 million digital televisions worldwide, a number bound to increase in the near future. Digital television is already a multibillion-dollar business. In fact, many countries have set a date for switching off analog broadcasting. Broadcasters are being forced to invest huge amounts of money to implement digital broadcasting, with the hope that this will bring in new revenue.

Digital television is an emerging method of television transmission. It is based on similar concepts to those employed in mobile digital telephone networks. Digital television works by transmitting television signals as compressed digital data rather than analog signals. Compressed digital data requires smaller bandwidth than its equivalent analog signal. This enables digital television to transmit a greater number of channels within the same transmission band than current television transmission systems using analog methods.

Digital technology can be used to transmit television broadcasts by cable, satellite, or as digital terrestrial television (DTT). Traditional television broadcasts are based on analog signal radio transmission along specific frequencies. Digital television uses the same approach as digital telephony; radio frequencies are used as the communications medium, and the signal is transmitted as discrete blocks of digital information, or data.

Using data compression technology, the discrete blocks of information can be compressed, freeing capacity along frequency channels. The use of compression and digital encoding means digital television can offer up to 500 different channels. In addition to greater channel capacity, digital television offers the opportunity to support additional technological developments, including support for higher resolution, wide-screen formats, and electronic program guides.

Of most interest to the target marketing professional, digital television delivered over cable or telephone can also provide support for interactive digital television, where the end user is involved in the program content and broadcast loop. This point is critical. Without a link back to the service provider, digital television offers wider choice but no means of direct feedback from the watcher. Nor does it offer a means of tracking each individual household's viewing patterns. In fact, digital television can be delivered through the normal TV aerial via DTT, but there must also be a hard-wire link between the viewer and service provider for interactivity to be possible, hence the importance of the imminent launch of interactive digital TV sets, which also offer this feedback link.

The digital TV can be delivered via satellite, cable, or terrestrial transmission systems. Europe, as well as several other areas of the world, has chosen to adopt the DVB standard. This is available in a number of compatible versions, including DVB-S for satellite transmission; DVB-C for cable transmission; and DVB-T, for terrestrial transmission. Europe has chosen digital terrestrial broadcasting systems with services that started in the later part of 1998. The digital satellite master antenna TV (SMATV) version—DVB-CS—has been adapted from DVB-C and DVB-S to serve community antenna installations such as blocks of apartments. Singapore has been conducting experiments with DVB-T, and Australia has plans to adopt this standard. India started using the DVB-T system in April 2001. The data broadcasting version of the standard—DVB-Data—allows a wide variety of different, fully interoperable data services to be implemented. DVB-S satellite transmission of data is much faster than traditional telecommunications methods.

Standards used in other parts of the world include the following:

- *DSS:* the proprietary digital satellite standard developed by Hughes Electronic Corp., used by its DirecTV service in the United States and Japan.

- *ISDB:* the variant of the DVB-T system developed by the Japanese Digital Broadcasting Experts Group (DiBEG). It differs in one key respect only, the use of an intermediate (software-driven) data segmentation system, whereby services such as radio, SDTV, HDTV and mobile TV can be flexibly allocated pieces of the overall service bandwidth. The Japanese ISDB system is anticipated to launch in Japan in 2003.

- *CableLabs:* also known as OpenCable—the digital cable system developed in the United States by the CableLabs project.

- *ATSC:* also known as DTV—the U.S. digital terrestrial broadcasting system, developed by the "Grand Alliance"—a consortium formed in the United States. ATSC uses the single carrier 8-VSB (vestigial side band) modulation technology, where DVB-T uses the multiple-carrier Coded Orthogonal Frequency Division Multiplexing (COFDM) system. The surround-sound audio system used by ATSC is the proprietary Dolby AC3 system.

Although both the ATSC/DTV and the DVB-T systems use MPEG-2, they are different in a number of ways [6]. There is no interoperability between the CableLabs and ATSC and DSS standards. In other words, MPEG-2 streams must be decoded and recoded for jumps from one of these systems to another.

1.6.1 DVB

DVB is divided in to three areas: digital terrestrial (DVB-T), digital satellite (DVB-S), and digital cable (DVB-C). The principles behind these are illustrated in Figure 1.5.

DTT combines the benefits of terrestrial broadcasting with digital technology. It is the most cost-effective method of reaching national coverage and delivering regional variations in television services. Furthermore, it is the only means of delivering portable and mobile television. DTT uses a transmission technique in order to achieve good quality, reliable, and efficient transmissions. The technique avoids interference caused by multipath propagation and resultant distortions. A digital terrestrial transmission channel is able to carry approximately 8 to 24 Mbps of information in a channel of 6 to 8 MHz of bandwidth. This will be less for mobile reception and large, single- frequency networks. To put this capacity in context, an SDTV program requires around 4 to 6 Mbps, depending on the content and quality required. The DTT channel would therefore be able to deliver at least three SDTV-equivalent digital services and possibly as many as six. In the United Kingdom, DTT is being carried on six channels. Once all analog transmissions have ceased and the UHF spectrum has been released, DTT could theoretically deliver between 30 and 40 services. A new spectrum-efficient possibility with DTT is that a single frequency could be used throughout an entire country to carry a multiplex.

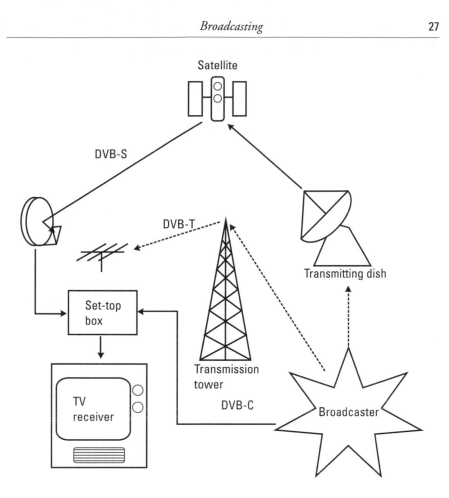

Figure 1.5 Digital broadcasting via satellite, cable, and terrestrial systems.

1.6.2 DVB-S

With advances in satellite communications technology, several DBS systems were put into operation during the late 1990s, each delivering about 200 television channels to subscribers with an 18-in (about 46 cm) dish antenna and a set-top converter. Video and audio signals, increasingly in digital format, originate at a ground station and are transmitted by microwave up to one or more geosynchronous satellites, which in turn transmit to individual antennas by microwave in a 12.2–12.7-GHz band. A subscriber's dish antenna is oriented in a fixed direction for best reception, at an elevation angle of approximately 10 to 50 degrees. Alternatively, it could be a steerable dish, however, some satellite broadcasters claim that if the dish is not permanently pointing to their satellite, software upgrades to the set-top box may not be possible. The satellite towards which the dish needs to point varies from country to country. For example, in Italy, the two commercial digital satellite broadcasters Stream and Telepiù, plus the public service operator RAI, all use Eutelsat's Hot Bird II satellite. In contrast, in the United Kingdom Sky Digital and the BBC use one of the Astra satellites.

The digital signal includes error-correcting codes to ensure receipt of acceptably reliable signals despite variations in signal level (fading) caused by rain. Encrypting the signals for those channels controls access to premium channels.

1.6.3   DVB-C

Digital cable offers a powerful and flexible range of services to consumers with more channel capacity, and therefore more choices, faster interactive services, and access to a large range of services already developed using Internet technology. Digital cable represents a significant evolution in cable technology. Digital compression technology allows up to 12 digital services to be carried in the space normally occupied by one analog channel, and it is not necessary to rebuild a cable system to enable delivery of the digital technology. Digital cable provides an exciting new television viewing experience without many of the disadvantages associated with satellite systems, such as expensive hardware, lack of local network stations, and the need to purchase additional hardware to view different programs on different television sets in the home.

Digital TV broadcasts via cable are only possible if the house is in an area that can be connected to a cable network that is broadcasting in a digital format. The cable operator would usually provide the set-top box and would also make a simple connection to the cable network. It has recently become possible to upgrade older cable systems to receive digital signals through a technique known as hybrid fiber coax (HFC) [7]. The modern HFC cable television (CATV) cable plant offers an excellent high-speed digital transmission medium between the CATV's head end and individual homes. Cable companies are laying high-capacity, high-speed digital backbone networks connecting their head ends and the outside world together.

Unlike satellite broadcasts, where the same data is transmitted over a wide area, cable has the advantage of being able to offer locally focused content, a feature that would appeal to the consumer.

1.6.4 Integrated Services Digital Broadcasting (ISDB)

Throughout the world, the trend is toward integrated digital broadcasting. Integrated services digital broadcasting (ISDB) is the standard being developed in Japan. In ISDB, analog video, audio, and test signals are digitized and these data, along with other digital data for control and value-added services, are formed into a transport streams (TS) using fixed-length packets. The conditional access is provided by multiplexing control signals into the data stream. It is used to address various services to each subscriber or to target programs to a specific group or customer. This leads to virtual channels for pay-per-view or VOD. Scrambling and decryption are provided at the receiver using hardware-embedded microprocessors or external smart cards. The digital video signal contains space during the line-blanking period that is not needed for picture information. It is possible to insert other data into the video signal into these spaces. There is a maximum capacity of insertion of 27-Mbps serial digital data stream, which can carry 8 stereo pairs of 3.072 Mbps [Audio Engineering Society/European Broadcasting Union (AES/EBU) standard AES-1992], or 16 audio channels or equivalent computer data. This data can be a software code, graphics, still picture, or any other value-added service.

ISDB will offer various broadcasting services expected to be available in the future, as shown in the diagram, integrated and edited flexibly on a single channel, and sent to households via various transmission media, including broadcasting satellites, terrestrial broadcasting channels, and optical fiber (Figure 1.6).

1.6.5 HDTV

Around the world, HDTV systems are being brought to the fore. Similar to the quality of sound gained from the compact disc (CD), HDTV sets a standard, in which picture quality approaches the clarity of 35-mm film. The standard for analog HDTV was defined in 1970 with 1125 lines scanning rate, line interlacing, 5:3 aspect ratio, and 60-Hz field frequency. The trials were conducted in Japan using Broadcast Satellite-2B (BS-2B) in 1989 [8], and experimental transmissions started in 1991. Japan has begun limited satellite transmissions but high receiver costs, combined with a lack of HDTV programs, are resulting in limited set penetration. In Europe and Japan, the emphasis is on analog satellite transmission. In the United States, the Japanese standard with 1125 lines/60 Hz was considered. However, there is the problem of terrestrial TV channels having a bandwidth of 6 MHz and priority being given to terrestrial program distribution over other transmission

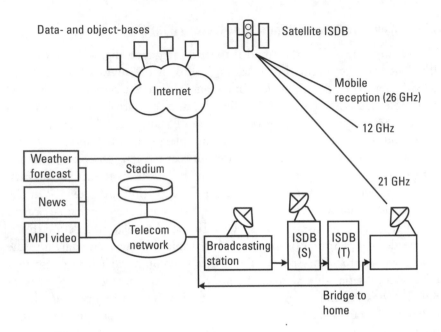

Figure 1.6 ISDB system.

media. Analog HDTV certainly cannot be implemented in the 6-MHz channel, so the United States is therefore focusing on digital terrestrial transmission systems using video compression.

Figure 1.7 shows the transmission chain for digital HDTV broadcasting. The high-quality video input is taken via a signal-processing unit before
applying the source encoder. The output data rate of 20 to 30 Mbps is subjected to forward error correction (FEC) channel coding. In the subsequent
multiplexer, the audio, sync data, and control signals are added to the coded
picture signal, whereon a gross data rate of 30 to 40 Mbps is obtained. The
channel coded data are modulated onto the IF carrier and packed into a channel bandwidth of 7 to 8 MHz (6 MHz in the United States). A mixer upcon-
verts to the RF carrier frequency in Band I to Band V (47 to 862 MHz) before
the signals are applied to the terrestrial transmitter and antenna.

1.6.6   Direct Broadcast Service

In the 1970s and 1980s, satellite became the chief means of long-distance
radio communications and facilitated worldwide TV program distribution in

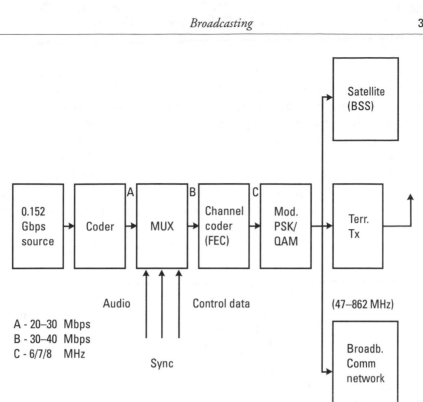

Figure 1.7 HDTV system diagram.

real time. In rural areas not serviced by terrestrial TV broadcast stations and cable TV systems, satellite signals made possible direct reception of TV from satellites by TV receive only (TVRO) receivers equipped with parabolic antennas with diameters between 2 and 5 meters. For almost two decades, TV program delivery by satellite was done in the 4 to 6 GHz and 11 to 12.4 GHz bands.

Development in advanced radio communications technologies and the offer of improved radio-based services paved the way for direct TV and audio broadcasts from satellites (also known as direct-to-home service). Today, three technologies have been aggressively developed: DBS, BSS-HDTV, and BSS-Sound.

DBS was introduced in the United States in late 1993 to deliver conventional television programming directly to the consumer using frequency of 12.2 to 12.7 GHz for the broadcasting satellite service (BSS). The DBS system consists of a small 18-inch satellite dish, a digital integrated receiver/decoder

(IRD) that separates each channel, decompresses, and translates the digital signal, and a remote control. The dish never has to track the satellite, so there is no waiting for the picture to come in [9]. DBS programming is distributed in the United States by three high-power HS-601 satellites (DBS-1, DBS-2, and DBS-3), colocated in geosynchronous orbit, 22,300 miles above the Earth at 101 degrees West longitude. DBS-1 comprises 16 Ku-band transponders at 120 watts, with DBS-2 and DBS-3 each configured to provide 8 transponders at 240 watts. DBS-1 delivers up to 60 channels of programming, and approximately 20 channels of programming are provided from United States Satellite Broadcasting (USSB). DBS-2 and DBS-3 are used exclusively to provide approximately 175 channels. Each of the transponders on the DBS-1 satellite can send more than 23 Mbps. The DBS-2 and DBS-3 satellites are even faster, at around 30 Mbps each.

DBS employs MPEG-2 technology, the world standard for digital broadcasts. DBS is fully digital and "forward compatible" so that consumers can take advantage of emerging technologies, such as interactive services, 16×9 wide-screen and HDTV (digital) broadcasts.

Programming comes from various content providers via satellite, fiber optic cable, or special digital tape. Most satellite-delivered programming (analog communication satellites used for program delivery) is immediately digitized and uplinked to the orbiting satellite. The DBS satellites retransmit the signal back down to every Earth station, or in other words, every little DSS receiver dish at subscribers' homes and business.

BSS-HDTV is meant for HDTV but due to the problem of large bandwidth required for analog HDTV, the service has not become operational. The World Administrative Radio Conference-92 (WARC-92) has allocated two frequencies for this service: 17.3 to 17.8 GHz for Region 2 (Europe) and 21.4 to 22 GHz for Regions 1 (America) and 3 (Asia). These new allocations will become effective April 1, 2007 [10].

Experiments are being made in labs for simultaneous reception of broadcasting from multiple satellites using multibeam receiving flat antenna (Figure 1.8).

BSS-Sound is for high-quality audio programming. BSS-Sound generally refers to the delivery of music, sports, news, and other programs directly to consumers' radio via satellite. WARC-92 adopted three different allocations for BSS-Sound: 1452 to 1492 MHz, 2310 to 2360 MHz, and 2535 to 2655 MHz. The service shall be received using portable radios. Such services will not be compatible with existing analog AM and FM radios and will require consumers to purchase new radios in order to enjoy this new broadcast service.

Figure 1.8 Multibeam flat antenna.

1.6.7 DTH System Description

During the last three years, the DTH satellite industry has come on strong worldwide. It has grown from a niche delivery mechanism into a mainstream business. The spread of subscription-based DTH satellite TV promises to enhance choices for many households. Using additional low-noise block (LNB) converters, which are used to amplify and convert satellite signals into frequencies sent to the receiver and are mounted at the focal point of the dish, viewers can gain access to a variety of channels via additional transponders. This feature of satellite television is the main advantage over its cable counterpart, targeting niche markets.

DTH has been prominent in many countries. In Britain, DTH became a success largely because cable television was not a popular medium. Although the United Kingdom started building its cable network in 1984, only 1.9 million out of 8.4 million television homes (22%) had enrolled by 1997. DTH was relatively more successful in the United States primarily because of the poor picture quality of cable TV and the pent-up demand for multichannel television among people not covered by cable. The DTH system is shown in Figure 1.9.

Encryption

All broadcast TV encryption systems have three primary components: an encoder, an authorization center, and a universe of low-cost, individually

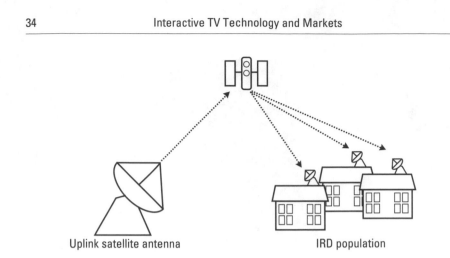

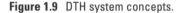

Uplink satellite antenna   IRD population

Figure 1.9 DTH system concepts.

addressable decoders. The encoder, which is located at the satellite uplink facility, converts the analog TV signal into an encrypted message that must be decoded before it can be displayed on a standard TV set. The encoder is also linked to a computerized authorization center that can process subscription orders and control all decoders in the system. Total control is achieved by assigning a unique address code to each subscriber's IRD. The authorization center uses this code to turn on or off any individual IRD or to control a selected group of set-top boxes. The authorization center, for example, can "black out" specific regions of the country when a particular programmer is transmitting a sporting event for which it does not own the national broadcast rights.

1.6.8   Costs Comparison of Various Delivery Systems

The largest cost component of digital television are the receivers—TV sets and set-top boxes. However, the costs of these are decreasing rapidly—at the rate of 20% per year in the case of set-top boxes. Manufacturers are developing more generic equipment to conform with international standards. There are costs of reception, that cannot be avoided and that are specific to delivery systems. Satellite reception requires a dish and an LNB. Cable reception requires physical connection to the cable network. In the case of multiset households, in-home distribution usually requires professional installation in cable and satellite homes. In the case of DTT, this advantage is being exploited by planning the transmission in order to achieve deep in-home reception (i.e., direct to the TV set-top aerial). The relative cost advantage that DTT has over other digital delivery systems clearly applies to the core area covered by the main transmitters. However, this depends on a country's topology and terrain conditions. The costs of satellite reception per household are greater than terrestrial. The exception is for reaching households in remote communities outside the reach of the main terrestrial transmitter network. Digital cable is only competitive in densely populated areas where wireless reception may be difficult.

1.6.9 Digital Radio

Digital radio is the transmission and reception of sound, which has been processed using technology comparable to that used in CD players. In sort, a digital radio transmitter processes sounds into patterns of numbers, or digits: hence the name digital radio. At the listening end, digital radio receivers provide a standard of sound quality that is significantly better than conventional analog radios, just as CDs sound better than LPs.

In broadcasting terms, bandwidth is the radio spectrum "space" required to deliver a given signal. Different types of signals require different bandwidths. For instance, current TV signals require 6 MHz of bandwidth per channel, FM requires 0.25 MHz, while AM radio only needs 0.01 MHz per channel. Unfortunately, what has stood in the way of digital radio is the amount of bandwidth required to transmit it—up to 1.5 MHz per stereo service. Digital radio has a number of varieties.

DAB uses L band or VHF band. The United Kingdom has gone for VHF band. Canada has gone for L band. American broadcasters are trying to develop a form of digital radio that will work within the existing bandwidths allocated to the AM and FM stations. Known as the "in-band on-channel" (IBOC) solution, it remains to be seen whether a satisfactory solution can be found for both AM and FM. For one thing, most countries want future digital services to be delivered by both satellite and terrestrial transmitters. Due to the transmission properties of AM and FM, the IBOC system is not practical for satellite use.

Digital radio mondiale (DRM) is an international consortium of broadcast industry organizations that have come together to develop a standard for digital broadcasting in the AM and shortwave bands below 30 MHz. Using short wave propagation via ionosphere, DRM can provide high-quality audio over a large area. A DRM receiver is shown in Figure 1.10. A DRM's aim is to bring affordable, digital quality sound and services to the world radio market. The DRM consortium was formally established in March 1998. Members

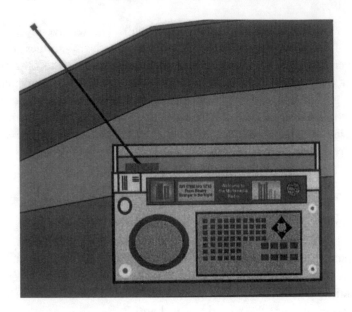

Figure 1.10 DRM receiver.

include leading broadcasters, network operators, transmitter and receiver manufacturers, research bodies, and regulatory groups from all corners of the globe. Membership continues to grow rapidly, as more organizations see the benefit of high-quality, wide-area coverage at low cost.

1.6.9.1 DAB

The prospect of sending higher quality audio to portable receivers is spurring the development of DAB. Europe has led the way with its Eureka 147 project [11]. The project was launched in 1986 and continued until 1994. The project's main emphasis was to carry out test DAB transmissions and consider new communications services that supplemented broadcast services (i.e., traffic information and management systems, data transmission to a specific group of users). DAB was carried out using an orthogonal frequency division multiplexing (OFDM) packet in which six stereo programs were transmitted with other data (Figure 1.11). The project led to detailed and comprehensive system definitions and a specification of standards [12].

Japan started DAB satellite services in 1992 for small fixed-panel antennas. The BBC started its introductory DAB service in September 1995

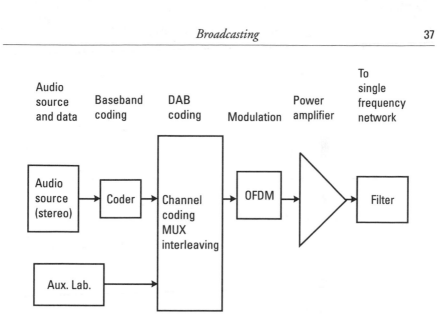

Figure 1.11 DAB transmission system.

in the 217.5–230-MHz frequency range. Broadcasters are investigating technologies to enable digital transmissions to share the same spectrum with existing FM and AM broadcasts. The advantages to the listener of the DAB system over FM are impressive. The sound quality is comparable to that of CDs and is relatively immune to multipath interference. Data such as lyrics and phone-in numbers can also be transmitted with the audio. Program labeling, graphics, and traffic messages are among the other enhanced features available. A block diagram of a DAB receiver is given in Figure 1.12. DAB is much more spectrum-efficient since a single frequency can be shared for the entire national terrestrial network.

There are a number of broadcasters broadcasting in a digital radio format in various European countries, including the United Kingdom and Germany. However, currently the main obstacle to this development is the high cost of radio receivers (about 700 Euros). It is also possible to plug a data card into a computer to receive digital radio. Digital radio not only offers high-quality sound, particularly when on the move, but also some limited multimedia as well.

1.6.9.2 Color Radio

An American company, GEODE Electronics, LLC, has introduced COLOR RADIO for radio data service (RDS) audiences. RDS will be discussed in

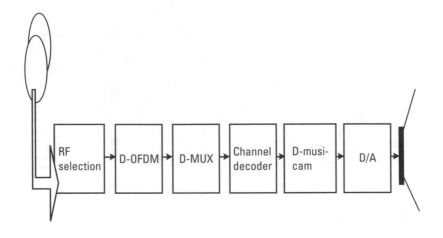

Figure 1.12 Block diagram of a DAB receiver.

Section 1.6.10. The company's proprietary radio technology provides listeners with text information on a color-graphics display. It has been estimated that out of 93 million radio receivers sold in the United States each year, 60 million are RDS COLOR RADIO candidates. COLOR RADIO is designed to allow a choice of colors and to display text in different font types and styles, with a variety of lines and borders. The color radio provides listeners with a series of screens displaying station and advertising information. GEODE claims that there are about 550 stations in the United States currently transmitting RDS signals. In addition to the receiver, GEODE has introduced what it calls RDS II to add features to the transmission end in the hopes that more stations will be attracted to radio data service [13].

1.6.9.3 Digital Multimedia Broadcasting

Digital multimedia broadcasting (DMB) provides video and hypertext data by DAB. It can transmit a television program at least in VHS quality (Figure 1.13) within a single DAB frequency block. This technique operates on the conventional DAB transmission technology (with COFDM modulation) and merely involves an exchange of useful contents.

Whereas, in the case of transmission in accordance with the analog PAL, SECAM, or NTSC standards, interference phenomena (echo images) practically prevent the reception of television programs in moving vehicles, a perfect picture can be received with DMB. Digitized television programs require uncompressed data rates of more than 200 Mbps. Since available

News $ traffic info.

DMB transmission of video and audio in a German metro train

Figure 1.13 DMB picture via DAB transmission. (*Source:* Robert Bosch Multimedia Systeme GmbH & Co.)

frequency resources are not sufficient for such capacities, the DAB compression technique employs high compression for transmission of television pictures. In the DAB system, the television signal is compressed to a data rate of 0.94 Mbps. Truly, the radio will not only be for the ears but also for the eye in the future (Figure 1.14). In combination with the audio information (MUSICAM[2]-reduced), additional data services, and error correction, a data rate of 1.152 Mbps is achieved, which is easily accommodated within a single DAB block of 1.5 MHz width [14].

The substantial advantage of DAB TV over analog techniques is the interference-free mobile reception. The principal potential applications for this system have been identified in the fields of program transmission into trains, long-distance buses, or other means of transport.

---

2. MUSICAM stands for masking pattern adapted universal sub-band integrated coding and multiplexing. It is a method for audio baseband coding.

Figure 1.14 Radio for the eye.

1.6.10 Data Services (Value-Added Services)

Digital sound and TV broadcasting also contain data, or value-added, services. The worlds of audio, video, and data are rapidly converging. Digital sound and video signals may be considered anonymous data streams. Efforts are ongoing to include one or more digital data channels along with broadcast signals. These data channels can be used for various value-added services. For example, video and data services can be used to obtain personal responses from television viewers and provide educational programming, conduct polls, download data, and order pay-per-view programming, services, and products. There is a demand for transparent networks using different ways of transmission via terrestrial transmitters, satellite, and cable with open interfaces to various services. DAB and DVB are becoming digital integrated broadcasting (DIB) [15]. Broadcasters are attempting to reach standards and implement a system under the name RDS. A few of the services that have been introduced recently or are being planned are discussed below.

1.6.10.1 RDS

RDS systems were introduced in 1984 to provide traffic information services for motorists in Germany [16, 17]. RDS uses a free space in the FM stereo multiplex signal at 57 kHz for a 1187.5 bps-wide data channel. The basic idea of RDS is to provide additional information for mobile reception. It can tell a receiver how to tune a station by format or call letters. Interactive paging is one of the popular services of the RDS [18]. Today, RDS has been adopted in almost all European countries, as well as in the United States, Japan, and India.

Radio data are transmitted in the form of a continuous, binary data stream with 1.1875 Kbps. In selecting the modulation carrier and type of modulation for the RDS signals, the existing occupancy in the stereo multiplex baseband has to be considered.

1.6.10.2 Data Radio Channel

The International Telecommunication Union–Radio Communication Assembly (ITU-RA) approved the data radio channel (DARC) system in October 1995 as a recommended system for large-capacity FM multiplex broadcasting to stationary and mobile receivers [19]. The DARC system, which has been developed by NHK (Japan Broadcasting Corporation), can provide a variety of data services for mobile receivers [20]. The system is used to provide traffic information, news, weather forecasts, and information services related to conventional FM programs. The system can also be applied to differential global positioning system (DGPS) and radio paging services. NHK started news and weather forecasts using DARC in March 1996.

The DARC system uses a specially developed modulation scheme called level-controlled minimum shift keying (LMSK), which maintains transmission quality and ensures compatibility with stereo sound signals, as well as RDS, by controlling the injection level in proportion to the stereo L-R signals. DARC uses the product code of the (272,190) shortened difference set cyclic code that can be decoded rather simply by using a majority logic circuit, which gives greater robustness in adverse environments. The data rate is 16 Kbps.

1.6.10.3 Advances in Data Broadcasting

Data broadcasting today involves delivering multimedia content directly to a computer. This can be done by installing a specific data card into the computer to receive the data. The data is converted into a format that can be used by the computer. As most commercial data broadcasting services deliver the data via satellite, the data card is connected to a satellite dish. Requests for specific data are sent via the "return path," which is a normal telephone line connected to the computer via a modem. A new generation of two-way satellite links using 90-cm dishes can also be used. It will also be possible for higher-speed return channel links to be available via cable modems if connected to cable networks.

Various European satellite operators, including Astra, Eutelsat, and Hispasat, have implemented satellite data broadcasting networks (DBN). The wide-area coverage of a single satellite footprint and the possible bit rates of more than 30-Mbps per transponder enable a typical CD-ROM to be transmitted to a whole continent in under three minutes. With more homes having PCs, and many households having direct access to satellite transmissions, the potential for using this technology is huge.

1.6.11 LDTV, SDTV, EDTV, and HDTV

It is worthwhile to know the quality expectations for the digital system. The digital system provides an increased signal-to-noise ratio of the order of 98 dB for 16-bit system [SNR = $(6.02 \times n) + 1.76$, where n is the number of bits per sample], as compared to 65 dB of analog audio [21]. There is no deterioration or multigeneration tape loss when it comes to long-term archiving. As discussed earlier in Section 1.4, for storage and transmission, the signal is compressed. Since the reception quality depends on the extent of compression, a concept of hierarchical quality structure has emerged for digital TV [22]. HDTV is for very high-quality viewing in a big hall. Extended definition TV (EDTV) is also high-quality viewing in the home. SDTV is the quality available today. Low definition TV (LDTV) is for mobile service in trains or cars. Table 1.10 shows the hierarchical quality structure for compressed digital TV signals.

1.7 Advent of Television and Its Social Aspects

As a point-to-multipoint communications system, broadcasting's traditional role was defined as "a medium for information, education, and entertainment" to a mass audience using sound or vision. Television was a major cultural socializing agent during the late 1950s. During the 1960s and 1970s, TV was considered the ruler of the mind of society. For 20 long years, until the late 1970s, TV was the unchallenged controller of the social psyche, since it was the only real provider of complete content for social integration.

Table 1.10
Hierarchical Quality Structure for Digital TV

Format	HDTV	EDTV	SDTV	LDTV
Quality	High	Enhanced	Standard	Limited
Comparable to	2X ITU-R 601	ITU-R 601	PAL/SECAM/NTSC	VHS
Data rate before compression	1 Gbps	432 Mbps	216 Mbps	108 Mbps
Data rate after source coding	30 Mbps	11 Mbps	4.5 Mbps	1.5 Mbps
Compression	MPEG-2	MPEG-2	MPEG-2	MPEG-1
Utility	Telepresence	Home viewing	Home viewing	Mobile TV

Television fulfilled an enormous role by providing viewers with common content and a common reference—essentially, a common ground for cultural involvement. The radio and TV created a sense of unity and community by allowing audiences to share the same information at the same time. TV was the most democratic of all the media, since it represented the people and was, by nature, a people's medium.

However, the 1990s witnessed rapid development of digital broadcasting, computer, and communication systems. The development of digital compression techniques threw open the prospects of multimedia and data broadcasting. The broadcast of sound, video, text, and data all into one stream became a reality. The transmission system changed from analog transmitters to a host of delivery media such as satellite, cable, Internet, and ATM/SDH networks. Several services could be provided over a common link. Broadcasters became information providers. The concept of global information infrastructure (GII) and national information infrastructure (NII) evolved with broadcast to become a component of the overall scheme.

Information is the key to all activities, and knowledge is the core of strength. Information can be used for production, as well as destruction, for good, as well as bad, purposes. A GII offers easy access to limitless information, creating new threats of culture invasion, programming of minds, and changing the way people behave. Defense personnel worldwide have realized the potential of information war and antiwar. The well-known three-dimensional space domain—land, sea, air, and space—is supplemented by the fourth domain: logic (human minds and will). Besides the conventional "hard kill" threat, the new emerging threat is "soft kill," characterized by unspecified and unquantified vulnerabilities of people's minds in a conflict. A hostile country can use this vulnerability to its advantage. Sustained programs of any kind can wipe away cultures, along with people's ethos and thinking process.

This leads to a very changed role of broadcasting in the new "information era" of GII NII. Broadcasters have to play their distinct, well-defined role of preventing people from the soft kill. Private channels may provide the programs, which may be commercially viable, but a national public broadcasting system will have to play a vital role. Market forces alone cannot be allowed to pave the way. Thus, the role of broadcasting will be very distinct in the future. Every television station—regardless of whether it is a public, private, educational, or commercial station—has an important role to play in providing the public with a collective consciousness. All TV stations provide common ground to a community. The public versus the private in television is key. The role of the "public" is to provide a sense that there is a "public." A private system consists of a group of consumers. Regulators will need to play a larger role and decide the roadmap of broadcasting of nations.

The first decade of the new century is witnessing the forging of a link of unprecedented magnitude and significance between the technological innovation process and economic and social organization. Countless innovations are combining to bring about a major upheaval in the organization of activities and relationships within society. A new "information society" is emerging in which the services provided by information and communications technologies (ICT) affect all human activities. With easier access to information, it is becoming increasingly easy to identify, evaluate, and compete with economic activities in all sectors. The pressure of the marketplace is spreading and growing, obliging businesses to exploit every opportunity available to increase productivity and efficiency. Social habits are changing drastically, and a new generation is emerging with new concepts, ideas, and thinking. Social responsibility is a matter of all those involved, including the government, in making this revolution happen. The information society is producing a significant acceleration of economic and social changes, and new and more flexible forms of knowledge societies are emerging that often require new forms of social protection.

In spite of the emerging technology in various fields of information, communication and entertainment (ICE), we still need television even today. There is no social unity today without television. In the late twentieth century, social unity was provided during the Renaissance through writing. In the early twenty-first century, TV will provide such unity because TV globalizes the culture. The economic impact of technological progress on growth and employment depends on TV's innovation process, which has become interactive. The linear model of innovation, with the innovative act being isolated, has, in today's world, been replaced by complex mechanisms: Innovation requires constant and organized interdependence between the upstream phases linked to technology and the downstream phases linked to the market. The means available to create, process, access, and transfer information is remodeling relationships in society. One of the most important current developments is the breathtaking expansion in how people can communicate and process information (sound, text, and images) in digital form. TV has been a generous medium; it has created an affluent society. People can afford not only the set but also the set's content. Television has not only generated an incredible acceleration of the economy it has greatly influenced the consumer society.

TV is a great medium for relaxing stress. Mentally, socially, physically, in every way, there is more human energy coming out of the incredible technological environment than we are putting into it. Through our constant use of telephones, computers, faxes, television, radio, and mail, we have become a population that requires an incredible amount of energy and flexibility, while also facing a lot of stress at the same time. We are in constant transition, pushed around by our technologies, our globalization, and the fact that the rest of the world is imploding on us. The couch potato is someone who is replacing the content of a working mind with the content of a mind that is having all the work done for him or her. TV is there to calm us.

Education for all and eradication of illiteracy is a motto of many developing and underdeveloped countries. There is a strong connection between alphabetic literacy, the ideals of democracy, and the constitution of an objective, rational body of references that are common to all. TV is a great medium for providing education in formal and nonformal sectors.

The idea of the global village is grounded in space. TV represents space—it shows space and the continuity of space, from Japan to Canada and from Canada to South America. There is a continuous space, which is represented by TV, and TV's role is to spot different elements of a single environment. That's the global village.

1.8 New Technology Frontiers in Broadcasting

Developments in digital coding, modulation, and compression have made the transmission of digital audio and digital video commonplace. These technologies are gradually making their way into the conventional broadcasting scene as digital audio and advanced TV (ATV). The potential exists for an enhancement to AM and FM broadcast stations, where the sound quality can equal or nearly equal that of CD-ROM technology. Also, HDTV is being developed in the United States, Europe, and Japan as a means of providing greatly improved picture quality to television viewers. The development of a successful HDTV system will provide the basis for revolutionary new video services to many homes, industries, science, and medical organizations. The implementation of HDTV is difficult and expensive for broadcasters, but it appears essential for broadcasters to find a way to upgrade their facilities to provide HDTV to consumers and remain competitive with the virtually certain introduction of HDTV by cable, VCRs, and DBS. International broadcast stations will experience improved spectrum efficiency with the planned single-sideband implementation and satellite-sound broadcasting potentially representing a supplemental delivery system to international audiences. The technology has reached the point where digital sound and picture are economical and have broadcast quality. Thus, broadcasting in recent times is undergoing a phenomenal change. It will be worthwhile to briefly review the technological trends toward the emerging broadcast system.

1.8.1 Studio Technology

The broadcast process consists of initial recording or shooting, post production, and distribution. Initial production involves shooting in studios, outside broadcast (OB) coverage. Postproduction consists of editing, layering, modification of picture contents and adding audio-video effects, while distribution consists of actual delivery via terrestrial transmitter, satellite, or cable (Figure 1.15). The limitation of analog technology with regard to the number of copies and layers is not the only drawback. The tape recorder needs time to play, record, wind, and rewind the tape to find the desired clip. The digital audio-video server and disk recording technology provide random access, ease of editing, and post-production. As such, the trend is toward a digital production system.

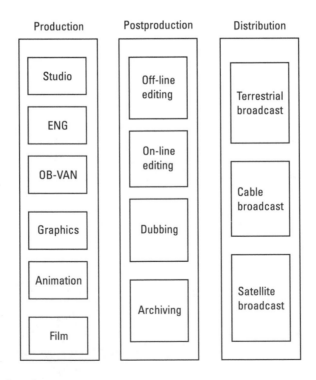

Figure 1.15 Broadcast processes.

In most digital broadcast installations, the system is built around a central signal distribution unit. This equipment receives various signals from internal and external resources. The internal resources are live or recorded material. The external resources are OB vans, databases, value-added signals, networks, etc. These are processed digitally to eliminate the generation loss.

Studio production is becoming more fascinating with the development and integration of computer system and computer languages. Research has continued on the use of digital technology, including computers for program production. Advanced broadcasting services friendly to the elderly and disabled, such as a system for captioning programs using automatic speech recognition and a speech rate conversion system, are under development (Figure 1.16). TV program-making language (TVML) is a language for making scripts, which are recognized by both humans and computers. Research is being carried out to produce computer graphics (CG) programs with a computer using this language (Figure 1.17). Research and development are also being carried out on technologies that automatically translate speech in a program into easily readable captions (Figure 1.18) [23].

1.8.2 Flexibility of Digital System

Pressure to produce content for markets beyond that of traditional 2D television, and to cut the costs of making programs, will force broadcasters to consider radical new approaches to program production. The production system will be independent of the delivery system. Audio-video professionals will

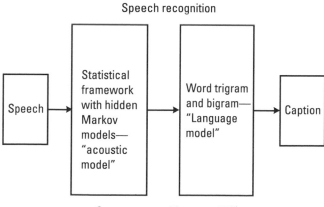

Figure 1.16 Automatic production of captioning from speech.

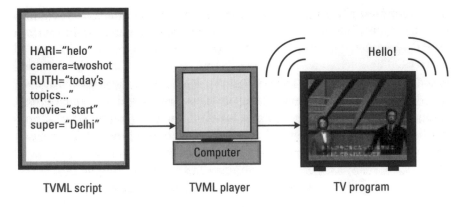

Figure 1.17 TVML.

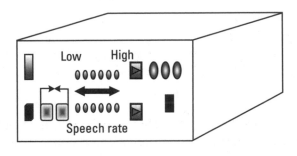

Figure 1.18 Portable speech rate conversion system.

create format-independent content that will be distributed via the Internet, CD-ROM and DVD-ROM, cable, DBS, and data broadcasts, to a new generation of digital media appliances, using emerging standards such as MPEG-4, MPEG-7, and MHEG. The content will have to be created in very different, but standard, formats to cater to the following needs.

1.8.2.1 Hypervideo

Hypervideo is digital video and hypertext, offering to its user and author the richness of multiple narratives, even multiple means of structuring narrative (or nonnarrative), combining digital video with a polyvocal, linked text.

1.8.2.2 Hypermedia

Multimedia is an amalgamation of audio, video, data, and text. Hypermedia is the union of two information processing technologies: hypertext and multimedia. Hypermedia is based on objects and time. Within a camera scene, there are multiple objects. The scene 00.01.16.12 (Figure 1.19), which is one camera shot, has got multiple objects, such as the diver, the vegetation, fungus, and the background scene. All of these objects can be directly linked to their metadata. The metadata is the detailed information about each object. This metadata can be accessed by clicking on the specific object. The link may also lead to more objects. Time-based, scenario-oriented hypermedia has been demonstrated in *VideoBook* [24, 25]. Here, multimedia content is specified within a nodal structure, and timer driven links are automatically activated to present the content, based on the time attributes. Hardman, et al. [26] have used timing to explicitly state the source and destination contexts when links are followed. MPEG-4 and MHEG groups are addressing this technology. Hypermedia affords immense potential for limitless information and education.

1.8.2.3 Video-to-Video Linking

Video-to-video linking was earlier demonstrated in the hypermedia journal *Elastic Charles* [27], developed at the Interactive Cinema Group (MIT Media

Figure 1.19 Object-based scene.

Lab). Digital video permits newer design and aesthetic solutions, such as those considered in the design of the Interactive Kon-Tiki Museum [28]. Rhythmic and temporal aspects achieve continuous integration in linking from video to text and video to video by exchanging basic qualities between the media types. Time dependence is added to text and spatial simultaneity to video. Narrative sequence is a path through a set of linked video scenes, dynamically assembled based on user interaction. Temporal links are the time-based reference between different video scenes, where a specific time in the source video triggers the playback of the destination video scene. Spatio-temporal links provide reference between different video scenes, where a specified spatial location in the source video triggers a different destination video at a specific point in time.

1.8.2.4 Multiple Perspective Interactive Video Architecture

Video provides a comprehensive visual record of environment activity over time. Multiple perspective interactive (MPI) video architecture provides the infrastructure for processing and analyzing multiple streams of video data [29]. A variety of mechanisms can be employed to acquire data about the "real world," which is then used to construct a model of this world for use in a "virtual" representation. Blending of "real" aspects in a virtual world is of great importance in applications such as telepresence systems, 3D digital television, and augmented reality scenarios [30]. MPI video provides the following:

- Content-based interactivity;
- Multiple perspectives;
- Virtual cameras, immersion;
- Scalable interactivity.

1.8.2.5 Immersive Video

Immersive video (IV) uses multiple live videos of an event captured from different perspectives to generate live videos of that event from any interactively specified viewpoint, thus creating a virtual camera. This new concept is an extension of MPI video [31, 32], which explores applications of visual information management techniques in ITV.

Let us assume that a scene is being recorded by multiple strategically located cameras. This scene may be a dance performance on a small stage, a football game in a large stadium, or a stage play that takes place in a theater. In IV, a remote viewer is able to walk through and observe this monitored environment from anywhere in the scene using virtual reality devices such as a head-mounted display or a boom (Figure 1.20). In addition, the view position may be tethered to a moving object in the scene, picked by the interactive viewer. Motion images, which are sent to the viewer's display, are created synthetically using computer vision and computer graphics techniques. To generate these images, IV uses a spatio-temporally realistic, 3D model of the entire environment. This environment model is built and continually updated by assimilating dynamic scene information extracted from each video stream. Special visualization techniques are then applied to create photo-realistic views of this model from the current position of the interactive viewer [33].

1.8.3   Multimedia Broadcast Services

Multimedia systems combine various information sources such as text, voice audio, video, graphics, and images. The potential applications are distance learning [34], multimedia mailing systems [35], collaborative work systems [36], multimedia communications systems [37], information and demand systems, [38] and multimedia broadcasting [39].

Figure 1.20   In IV, a remote viewer is able to walk through and observe a scene from anywhere using virtual reality devices such as a head-mounted display or a boom.

Multimedia broadcasting is defined as a suite of interactive digital services that combine to provide high quality, multichannels, and mobile reception. This is capable of providing digital audio, video, 3D TV, data, near-VOD (NVOD), computer software delivery, multichannel transmission, pay per view, value-added services such as RDS, paging, telemusic, teleshopping, and teletext. Multimedia broadcasting is based on the concept that the digital coding of audio, video, and other signal results in a data stream. As such, the difference between audio, video, and computer data ceases to exist after digital conversion. It is in this context that multimedia broadcasting assumes significance. Once the channel is digitized, more and more services, including audio, video, voice, graphics, animation, and computer data can be multiplexed [40] and put on a common delivery system. This transport system could be fiber-optic cable, terrestrial broadcast transmitters or a satellite in a Ku- or Ka-band capable of delivering signals directly to homes using a 50-cm dish antenna.

1.8.4   VOD

VOD is a service where a user uses a remote control to select the kind of video to watch and the time of its viewing [41–44]. Depending on the level of interactivity that the user enjoys, the VOD system can be categorized as below:

- *Interactive VOD (IVOD)*. The user has full virtual VCR capabilities, including fast forward, reverse, freeze, and so on.
- *Staggered VOD (SVOD)*. Movies with staggered start times allow individual viewers to choose their viewing time and "pause and restart" movies as they wish.
- *NVOD*. In NVOD, the user is a passive participant and has no control over the session except in choosing which program to watch.

Fully interactive VOD service is based on video servers. The multimedia is placed on this server, and is retrieved by the subscriber [45].

1.8.5   Information Superhighway

With its growing ease of use and burgeoning popularity, the Internet is fast becoming the all-purpose information superhighway [46]. The predominant technologies to broadcast on the Internet use buffering, codec (compression/

decompression), and streaming technologies. Buffering is provided to make up for transmission delays. By allocating a portion of memory to store a few packets, usually a dozen or so of audio-video information, the player always finds data to play from the buffer rather than waiting for receipt of data from the server. Codec technologies compress the data using compression algorithms at the server end and then decompress the data at the receiving end. Stream technology allows for real-time repositioning within a file, as well as playing files as they are downloaded. The main services to use on the Internet are described below.

1.8.5.1 Audio on the Internet

Until the advent of several lossy compression techniques, the large size of audio files coupled with the Internet's bandwidth limitations made it impossible to use the Web to efficiently and reliably access large volumes of archived audio content. This was because an entire audio file had to be downloaded to the user's machine before playback could begin. Buffering, codec and stream technologies have made it possible to deliver sounds on the Web, even when using a 14.4-Kbps modem, in real time with controls for rewind, forward, pause, and playback. The use of audio stream (continuous-delivery) technology permits playing a single audio packet on receipt. Since the transfer communications is bidirectional, the player can request the server to send a specific audio packet [47].

Streaming audio technologies are designed to overcome the Web's limited bandwidth: a 14.4–28.8-Kbps modem or 128-Kbps ISDN connection. A 14.4-Kbps modem has a throughput capacity of 1.8 Kbps, compared with the requirement of 176 Kbps of CD-quality audio (97 times the capacity of a 14.4-Kbps modem). For this reason, all the streaming audio technologies compress the data drastically to match the throughput of the Internet connection. While CD quality audio requires a compression of 97:1, several audio codecs start with lower quality; for example, 8-kHz, 16-bit audio requires a compression of only 8:1.

1.8.5.2 Video on the Internet

It is time to view the television industry in an entirely new light. Noam talks about the three-television revolutions [48]. The first television revolution was the introduction of commercial broadcast television about 50 years ago. This was called the "Privileged Television" revolution because it was characterized by relatively few broadcast channels. The basic operational model in the Privileged Television era was that each station transmission facility broadcast a single program service. The second revolution was "Multichannel Television." In this period, cable television and satellite services offered multiple program services over the same facilities. Cable audience research reveals that primary reasons for subscribing to cable are to improve signal quality and have more choice. The third revolution, according to Noam, is the "Cyber Television" era. More channels do not characterize Cyber Television. In fact, with the development and availability of devices such as video servers, switchers, routers, high-capacity telecommunications links, agents, and PC- or TV-type terminals, the whole paradigm of a "channel" is obsolete. Rather than selecting among diverse channel offerings, users can connect to content directly, so in a sense there is only one channel—the Me Channel. As the content layer becomes ever more separated from the distribution layer, the true impact of convergence will be realized.

As a medium, video is much more demanding than audio, both technically and aesthetically. Compressing television quality video, whose original bandwidth is about 27 Mbps, to a usable 28.8-Kbps modem, requires an astounding 7500:1 compression ratio. This extreme compression, achievable only by lossy techniques, causes tremendous distortion in the form of pixelation, blockiness, and gross artifacts. Using a 64 Kbps single-line ISDN or a 128 Kbps dual-line ISDN greatly enhances the quality of the video. A high bandwidth network or T1 connection can play a stored file at its full frame rate [49].

1.8.5.3 Music Archives on Web

The Web has opened exciting new possibilities for the music industry. Many recording artists, recording studios, and record companies now operate their own Web sites. Several radio stations have gone on-line and are broadcasting programs over the Web. In the near future, Web-connected fans will be able to log on from the comfort of their own room to hear their favorite musicians perform [50].

1.8.5.4 Virtual Radio

Virtual radio is the nonstop user-definable music broadcast on the Internet that brings the latest in new music. Virtual radio offers a wide variety of choices. Users can listen to a song on-line or download it to a local machine. This is not a sample song but rather a radio-quality broadcast of the entire cut, many times right off the bandmaster's DAT. Each "cyber tune" (a fancy name given to the music being broadcast on the Internet) page contains band information, a description of the music, and images of the band. Virtual radio is the new way to be exposed to today's music. Many broadcasting organizations such as the British Broadcasting Corporation (BBC), Malaysia

Radio and TV (MRTV), American Broadcasting Corporation (ABC) and others are producing programs and broadcasting them. The beauty is that an individual can also be a broadcaster by renting time on a Web site [51, 52].

1.8.5.5   Digital Music

The proliferation of the Internet and growing consumer demand for home computers has made the direct digital downloading of music a realistic commercial option. Several companies have already begun selling music directly to consumers. Technology that allows the secure downloading of music files direct to PCs has already been developed. The development of a music file format called MP3 that compresses sound files to send over the Internet has led to this explosion. Few companies used copyrighted music. However, a vast majority of companies sell unlicensed and illegal music. The MP3 files are very popular: behind "sex," it is the second most commonly used key term processed by search engines. Some legitimacy has been bestowed on MP3, with Lycos establishing a special MP3 search page, and the development of portable MP3 music players such as Diamond Multimedia's Rio player, which can be loaded with more than an hour's worth of MP3 files from a PC. Until recently, five major record companies—BMG, EMI, Sony, Universal and Warner—that account for 80% of the world's music have been reluctant to make their products available due to piracy fears and the lack of copyright security [53].

1.8.5.6   Secure Digital Music Initiative

The Secure Digital Music Initiative (SDMI), announced in December 1998, is the industry's definitive response to the growing popularity and acceptance of direct downloading of music as a realistic delivery method. It is also a riposte to digital pirates. SDMI includes both music and technology companies and has the aim of developing an open interoperable architecture and specification for digital music security. SDMI's mission statement describes its objective as one that will "enable consumers to conveniently access music in all forms and allow artists and recording companies to build successful businesses in their chosen areas." SDMI has the backing of the five major record companies and key technology players in the Internet, electronics, software, and telecommunications business.

1.8.6   Interactive 3D TV Network

A multimedia communication network that demonstrated the feasibility of 3D television transmission through satellite was demonstrated at the International

Television Symposium in Montreux, Switzerland, in 1995 [54]. The bit rate was 30 Mbps and the satellite was operated in the 20–30-GHz range. Interactivity was supplied via a reverse feed modem, allowing a robot arm in the Darmstadt studio to be controlled from the auditorium in Montreux. The demonstration aimed to show the viability of a viewer-controlled robot device via the Internet to select a program or interact with the broadcast.

1.8.7   Provisional Programming Service

The provisional programming service creates a new temporary programming channel while maintaining regular programming but reducing the bit rate. For example, when the time slot for a live HDTV relay of a match is extended due to the game, this service will reduce the HDTV bit rate to create an SDTV channel for the continuation of the match. In this case, the picture quality of the HDTV program will be of SDTV. However, the next scheduled program can begin on the HDTV channel without any delay (Figure 1.21). Selecting the regular HDTV channel or the SDTV channel is the viewer's choice. Viewers will be notified by an announcement or closed-captioning during the HDTV program they are currently watching. As the program on the provisional channel ends, the channel will automatically switch back to regular HDTV programming. When this service is provided, the HDTV and SDTV picture quality will depend on the bit rate allocation of each channel.

1.8.8   Multiview TV

Multiview TV is an application that divides one regular HDTV channel into a maximum of three SDTV channels (Figure 1.22). This is to broadcast program-related information on multiple channels simultaneously with the main program. The main program starts on the general channel, allowing the viewer to freely switch between the main and subchannels during the program. As the multiview program ends, the screen will automatically return to the next HDTV program on the main channel.

The multiview TV function can be applied to theater relays and dramas, as well as sports programs. This function has the possibility of expanding the viewer's selections and providing new information.

1.8.9   Hierarchical Modulation Service

Satellite digital broadcasting is affected by heavy rain (rainfall attenuation) causing abrupt audio-video blackouts (video: freezing; sound: muting).

Fax Back Card

Free information on other Artech House books...

To ensure that you stay up-to-date on the newest Artech House books of interest to you, fill in the address details below so that we may add you to our mailing list. Please fax or mail it to one of the locations listed on the other side of this card.

Name:

Position:

Company:

Address:

E-mail:

(If you would like to receive new title information by e-mail.)

Please indicate your areas of interest:

- ☐ Telecommunications
  Wireless
  Networking

- ☐ Computing
  Software Engineering
  Security/E-Commerce
  Database Management

- ☐ Microwave

- ☐ Radar
  Remote Sensing
  Electronic Defense

- ☐ Antennas & Propagation

- ☐ Signal Processing

- ☐ Optoelectronics

- ☐ Technology Management/
  Professional Development

AH Artech House Publishers BOSTON • LONDON

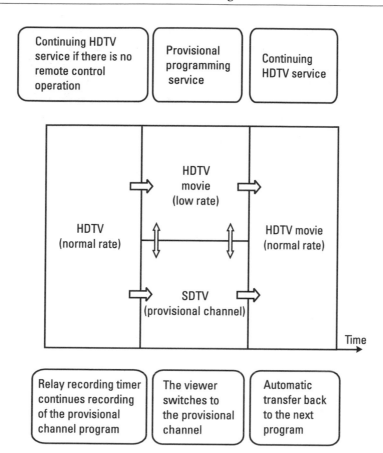

Figure 1.21 Concept of multiprogramming.

Broadcasting a program using two simultaneous modulations—normal modulation plus a second modulation (low hierarchical) that is resistant to interference by multipaths along with the regular modulation (high hierarchical)—will enable the transmission of information during rainfalls (Figure 1.23). Although this low hierarchical modulation is resistant to rain, the amount of information that can be transmitted decreases, which means that only necessary information will be sent on this system. A program that is normally in an HDTV video form could be broadcast with reduced sharpness, as well as reduced size, or as HDTV still images. A receiver will evaluate reception quality and automatically (or manually) switch between these hierarchies.

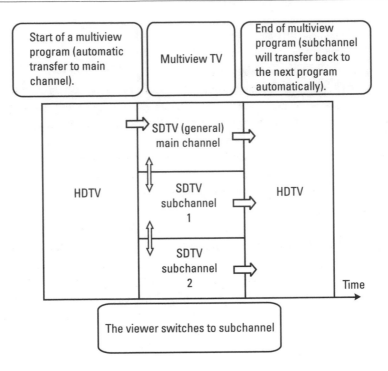

Figure 1.22 Concept of multiviewing.

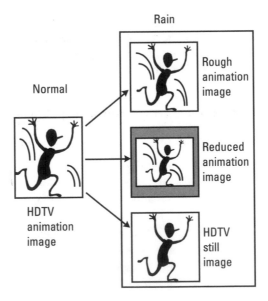

Figure 1.23 Hierarchical modulation service.

Since sound data is smaller, sound service can be sent by means of low hierarchical transmission. However, this low hierarchical transmission is not capable of transmitting numerous data broadcasting programs, resulting in the broadcast of only important programs, such as emergency reports.

1.8.10 Emergency Alert Broadcasting

A signal of emergency alert broadcasting is able to activate a turned-off (stand-by) receiver for emergency alert broadcasting. Further, if emergency broadcast starts while watching a program, the screen will automatically switch to the emergency programming. Research is being conducted on a system that will translate date information into Braille or speech synthesis for conditions such as disasters (Figure 1.24).

1.8.11 Home Servers for Integrated Services TV

TV receivers coupled with home servers (Figure 1.25) can store the program with or without the intervention of the viewer. This type of TV will be able to realize "anytime" functions.

1.9 Broadcast to Personal Computers

Broadcasting is an efficient way to distribute information. Existing broadcast networks already reach an enormous national and international audience. Broadcasts also reach a growing number of households internationally as broadcast satellite networks continue to proliferate. Broadcast to computers provides a flexible and cost-effective path to television of the future. Not only do they have a lot to offer immediately, but they also support low-cost

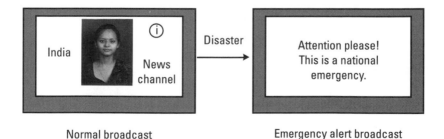

Normal broadcast　　　　　　　　　Emergency alert broadcast

Figure 1.24 Emergency alert broadcasting system.

Figure 1.25 Integrated servers with home TV.

incremental steps to higher television resolutions, growing back-channel bandwidth, increasing interactivity, and new multimedia forms of television. For viewers and content producers alike, broadcast to computers provide painless interim solutions at every stage. Rather than becoming obsolete when new technology becomes available, they are designed to incorporate technological advances smoothly.

Personal computers can receive video and digital data from virtually any broadcast source, including satellite, cable, and terrestrial television networks. Any computer equipped with enabling hardware and software can receive programs delivered through high-speed broadcast channels. This is an evolutionary step in the merger of computers and television.

IP multicast and other standard technologies are used to let a PC receive unidirectional digital and analog transmissions over any kind of television or computer network. Broadcast-enhanced computers can act as "clients" in a standard client-server computing model. These clients serve as data tuners that receive and process streams transmitted to them by broadcast "servers."

Broadcast to computers uses the existing IP standard for broadcasting data. Over the Internet and other computer networks that make use of IP, broadcasts take the form of IP multicasts sent to many recipients at once, in contrast to the usual unicasts that are directed to a single recipient. In a corporate context, multicasting can greatly reduce network traffic over Intranets when compared to unicasting the same data to the same recipients. A broadcast client is a perfect client for corporate multicasts because of its high-bandwidth capabilities and because broadcast to computers handles all broadcast data as standard IP multicasts.

Figure 1.26 illustrates client computers receiving broadcast data over different kinds of television network.

1.9.1 Broadcast Push Model

Broadcasting is an efficient way to deliver large amounts of information. Data transmitted over a computer or broadcast network using a unidirectional "push" model can reach a large number of people much more efficiently than the bidirectional "pull" model used on most computer networks today.

Computer networks today, including the Internet, generally use a pull model to transmit data. In a pull model, a client sends a request for specific

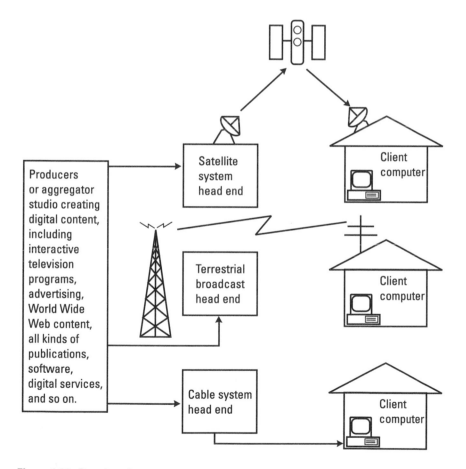

Figure 1.26 Broadcasting to computer.

data to a server. The server processes the request and then sends back the requested information. In this model, clients "pull" information from the server. The process breaks down when too many clients request information at once. This breakdown occurs because even very powerful servers can handle only a limited number of requests at a time, and they must send a separate response to every request.

In a push model, the server broadcasts a large amount of information onto the network on its own schedule, without waiting for requests. The clients scan incoming information and save what they are interested in, while discarding the rest. In this model, one server transmission can service an unlimited number of clients at once. In cases where many people need the same information, this push model is a much more efficient use of network bandwidth than the pull model.

The combination of a true push model and the very high bandwidths of broadcast channels allow broadcasters to deliver large quantities of data to customers conveniently. Such data can include video, audio, high-resolution images, large aggregated blocks of Web pages, databases, software, and data in other formats. This kind of data is generally too large to send or receive over telephone connections on a regular basis, even with the fastest modems. The advantage of the push model is that broadcasters can deliver this data to a client automatically, in the background, without the customer ever having to dial in, tune in, or download anything.

High-speed digital data streams with capacities of a few kilobits per second to many, many megabits per second will be available on broadcast clients. At 1.2 Mbps, a channel can transfer over 10 GB every 24 hours, while 6 Mbps translates into more than 60 GB of information per day. By contrast, a CD today holds about two-thirds of a gigabyte. In future versions of broadcast architecture, a variety of 30-Mbps channels may each deliver over 300 GB per day.

1.9.2 Client

As a receiver of television, radio, and data broadcasts, a PC has many advantages over older technologies. The combination of television and computers creates a new medium that not only provides information and entertainment for users, but also offers many new opportunities for advertisers, broadcasters, network operators, all types of publishers, and hardware and software manufacturers.

A broadcast client can be equipped with appropriate receiver cards and supporting software to receive broadcasts in virtually any format from almost any source. The digital or analog broadcasts of video, audio, or binary data stream by satellite, cable, conventional terrestrial antenna, or over a computer network can be received by the client using the Microsoft® Windows® 98 operating system. Almost all the technology and infrastructure needed to create broadcast clients is currently in place. In particular, broadcast clients rely on broadcast networks and a phone-line back channel that are inexpensively and reliably available in virtually every home today.

1.10 Super Surround Audio-Visual System

Research is being carried out on a next-generation broadcasting system that gives the audience the powerful impression and realistic sensation of being on site using a 3D surround image and sound system (Figure 1.27) [55].

1.11 Tools and Techniques

A variety of tools and techniques are available to cater to the above needs, albeit most are in infancy. There are several algorithms, the tool for writing

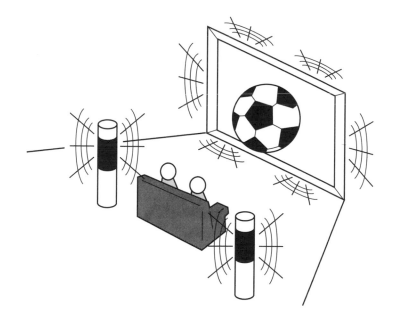

Figure 1. 27 3D surround image and sound system.

codes, available to implement these tools and techniques. A few of these tools and techniques are described below.

CORBA

Language-independent and platform-independent architecture by the Object Management Group (OMG) provides a standard mechanism for defining interfaces between components, as well as some tools to facilitate the implementation of those interfaces using the developer's choice of languages [56]. This means that a client written in C++ can communicate with a server written in Java, which in turn can communicate with a server written in COBOL, and so on. A single user interface can query all servers.

ORB

ORB is a software component whose purpose is to facilitate communications between objects.

JAVA

The JAVA applets in a user's set-top box maintains that user's profile, which helps the transmitting sites push hypermedia programs as per the user's taste [57]. We shall discuss Java technology for ITV in Section 3.6.

Server Push Technology

Server push allows delivery of multimedia content without requiring the customer to request or "pull" the information. With this mechanism, the server sends down a chunk of data. The browser displays the content and leaves the connection open to receive more data for a fixed time or until the client interrupts the connection.

Virtual Human Agent

A "multimodal interactive interface"—a virtual human agent—is a graphical user interface (GUI) system that uses human modes of communications, such as sight, hearing, talking, facial expression, and gestures [58].

Video Parsing Scene Detection

Video parsing is the process of detecting scene changes or the boundaries between camera shots in a video stream. Video parsing is done by temporal segmentation and content abstraction [59]. Video parsing or scene/shot change detection in a video stream is commonly used to extract key frames in a video stream. Key frames (the first frame of every shot) are extracted using the properties of color, texture, shape, and edge features. These key frames are then used for rapid video browsing and automatic annotation and indexing of video streams to support content-based query access to large video databases. The desired scene or video clip is extracted by using one of the following methods:

- Key-frame-to-key-frame extraction;
- Match/search the content from key frame;
- Query by visual templates—predefined templates (sky, sea, beach, forest) to formulate query.

Color has excellent discrimination power in image retrieval since it is very rare that two images, if they are totally different objects, have similar color. Color can be represented by a histogram of the distribution of color components. Average brightness, color moments (including mean), and dominant color are used for shot detection. In most images, a small number of color ranges capture the majority of pixels. These colors are used to construct an approximate representation of color distribution. These dominant colors can be easily identified from the color histograms of key frames.

Motion is the major indicator of content change. Dominant motion components resulting from camera operations and large moving objects are important sources of information for video parsing, for example a zoom shot is abstracted by the first, last, and middle frame.

Key frame retrieval using texture is done through Tamura features and simultaneous autoregressive (SAR) models. Tamura features include coarseness (granularity—changes in average over window), contrast (distribution of pixel intensities), and directionality (direction of gradients at all pixels). The SAR model provides a description (mean and variance) of each pixel in terms of its neighboring pixel. Edges (derived by Sobel filter) provide good cue for content. Shapes are derived by spatial segmentation algorithm [60].

View Processor

Multiple cameras are used to capture views from different angles. The view processor allows the user to select a viewing angle like a virtual camera.

HyTime

HyTime defines an extensive metalanguage for hypermedia documents, including general representations for links and anchors, a framework for positioning and projecting arbitrary objects in time and space, and a structured document query language [61]. The purpose of HyTime is to standardize some facilities for all the applications. In particular, it standardizes those facilities having to do with addressing portions of hypermedia documents and their component multimedia information objects, including the linking, alignment, and synchronization that standardized addressing makes possible.

The HyTime coordinate addressing is a generalization of the time model originally developed for another standards project, ISO/IEC 10743, the standard music description language (SDML). SDML is now an application of HyTime. It is intended to foster the growth of applications that bring music into the information processing world and apply information processing technology to the musical domain (for example, music publishing using modern text processing technology, including the integration of music with text and graphics).

Actor Motion Capture

An actor motion capture system allows simultaneous production of both conventional TV programs and 3D-interactive program content. The motion and facial expression of actors is captured in real time. Current systems based on reflective markers or magnetic trackers interfere with actors' freedom of movement and with the normal process of TV production. An ideal system would work with multiple actors over a large space, be completely noninvasive, and work in real time without operator intervention. For conventional TV, images of real actors in the studio set will be used, similar to today's virtual production. For 3D content that retains the appearance of 2D production, 3D models of the actors will be created by texture-mapping the studio camera images onto 3D models. This may require additional cameras in the studio, or other specialized 3D capture devices, but these could coexist with conventional TV studio equipment. Relatively crude models may suffice for some applications, particularly those where the possible viewpoints in 3D space are restricted. In its simplest form, each actor could be texture mapped onto a flat plane normal to the camera's direction of view, at a depth corresponding to the depth of the actor. This would allow a "2½D" scene to be formed and enable virtual foreground objects to obscure real objects without the need for separate mask signals and mixers [62]. For 3D content with virtual actors, information relating specifically to a 3D face and body model may be derived. This may be used to animate a virtual human or other virtual character. With the 3D information expressed in terms of body model parameters, it becomes easier to edit the movements in postproduction or to control the motion of the character by other means, for example, under software control in an interactive program.

Authoring Tools   Much software already exists to edit motion capture data and create 3D programs. Further development of such software will be needed to support open standards for data, such as MPEG-4 and MPEG-7. Authoring tools must also allow the creation of content in multiple forms, including both conventional 2D and 3D, with degrees of user interaction ranging from none, through viewpoint selection, to fully interactive or immersive. A further development will support the creation of shared virtual environments to allow multiple users to experience and inhabit a shared virtual world. Examples of collaboration technologies include video conferencing such as Picture Tel, MBONE, and Tango [63]. It may be possible to extend the use of MPEG-4 to such environments [64].

1.12 Analog Cutoff

Analog broadcasts will continue to run in parallel with new digital services, but eventually these will be switched off. In the intervening period, "simulcast" mode will be resorted. Currently, this is due to happen in Europe between 2008 and 2013, although the industry is very reluctant to set a date. A realistic date is 2010, although 2006 has been designated the cutoff date in the United States. Cutoff will not be sooner, as the process of converting the public to a new technology will be long and tiresome and will not happen as smoothly as broadcasters would like. In the United Kingdom, only 25% of United Kingdom television households have accepted cable and satellite. There will be a small proportion of the population that will not want to be forced into buying either a new television or set-top box. For this reason, analog broadcasts are needed to run in parallel with new digital programming. The consequence of this is that content providers and broadcasters will have to provide and support television on two different platforms until such a time that the majority of households have adopted digital technology. The simulcast of the broadcast system in analog and digital mode is going to put additional burdens on broadcasters.

1.13 Development of ITV

Television is considered [65] central to most people's domestic life and to the cultural, social, political, and consumer awareness for billions of people throughout the world. But it has been primarily a medium that is passively viewed when the broadcaster wishes to make the programs available. Although the videotape recorder has allowed viewers to record program and watch them when they wish to, this does not provide real interactivity. During the early to mid-1990s, the term "interactive TV" was used to describe feedback from viewers during live TV programs. Interactivity has been in the form of e-mail, telephone calls, or video linkups via ISDN to pre-senters in the studio. The use of digital broadcasting is claimed to offer new possibilities.

The 1990s witnessed many expensive proprietary interactive television projects. Although many of these projects failed, they led to many other spin-off business opportunities that are now starting to become commercially available. For example, VOD and radio-on-demand services through existing telephone lines to the home have been proven to be uneconomic due to the lack of demand for the limited number of videos and audio. Even accessing the Internet with POTS has been unsatisfactory, and WWW is often referred to as "World Wide Wait." But with the advent of DSL, users can be pro-vided with higher speed access to the Internet. In fact, this development could run in parallel and is very likely to compete with the next stage in the development of interactive TV.

With the development of digital TV, interactivity is now being under-stood to mean enabling millions of TV viewers to interact with programs and services. Interactivity can be divided into two types:

- Enhanced TV;
- Interactive services.

The method of accessing interactive services is through the remote con-trol or a remote keyboard, sending an infrared signal to the set-top box. The set-top box then displays that information on the TV if it is already stored in the set-top box or from regularly updated data rebroadcast on a carousel basis, or the set-top box may request specific information via the "return" channel through an existing telephone line. This information may come back either via the broadcast link or the telephone line, depending on what service is requested. Alternatively, with advanced cable networks, the request for data is sent via the cable system, which itself is a two-way broadband system.

Enhanced TV means that viewers can do more than passively watch a TV program. For example, they could do the following:

- Watch a travel program and click on an information icon that provides more detail in the form of text and graphics about the location of activities;

- Place an order or request further information about a product while watching an advertisement;

- Watch a football program and be able to view the match from a number of different camera angles with a split screen;

- Compete with studio competitors in a quiz show and have the results displayed at the end of the program;

- Access a directory of words or phrases during a language-learning program.

Interactive services involve accessing information independent of a TV channel. For example, such services could allow viewers to do the following:

- Access an electronic program guide (EPG) to select channels through a menu system updated regularly to ensure the program timing is up to date (the EPG could also provide additional information about the programs);

- Select information from an enhanced teletext-type service [66], which will display a screen in the form of high-quality graphics, various fonts, pictures, sound, and video clips;

- Answer multiple-choice questions and learn whether the answers are correct or not;

- Access an electronic encyclopedia requesting information about a particular topic;

- Send and receive e-mail;

- Shop on-line;

- Use an on-line banking service;

- Book airline tickets or other services;

- Access the Web;

- Order VOD services;

- Play interactive games.

Internet Access Via TV

The late 1990s have seen developments toward accessing the Web via the TV, either using a set-top box or a dedicated terminal. A number of manufacturers have developed and are planning to market a device that provides access to the Internet directly via an ordinary television set. The main thrust has initially been through WebTV [67] from Microsoft. The NetBox has been specially developed to offer Internet access via an ordinary TV set. It is based on software and hardware architecture developed by NetGem [68]. A key issue is the quality of the fonts and graphics that can be displayed on an ordinary TV set, which will be of a lower resolution than a computer monitor. The price of the device will also be critical, as it will need to be significantly lower than a home PC, which continues to drop in price. It is possible that these Internet TV devices will find a niche market—and a socially important one at that—for people who cannot afford to buy a home PC by providing them with access to the opportunities, including learning, of the Information Society.

1.14 T-Commerce

In earlier times, life was simple. I needed vegetables and you needed milk. I allowed you to milk my cow in exchange for a bag of vegetables from your field. This was plain and simple barter commerce. However, a direct exchange of goods or services was not feasible all of the time and cash was introduced. Mankind changed from "b-commerce" to "c-commerce." With global trade, cash exchanges became difficult, and an era of electronic commerce began. However, e-commerce requires computers and dedicated attention. Commerce through television, known as "t-commerce," has been found to be an exciting proposition. Since people love watching television, they also want to buy on impulse. T-commerce fulfills this need. For example, a woman may be watching a film. She likes a dress she sees, points to the dress, clicks a few buttons on the remote control, and immediately buys the dress. Television commerce in simple terms is the convergence of e-commerce technologies and interactive television.

Electronic commerce changed the nature of shopping forever. Dot-coms infiltrated every market sector from groceries to electronics to cars. Now, we will see the arrival of t-commerce, which merges the ease of television's use with the immediacy and e-commerce capabilities of the PC. Widespread penetration of digital television will lead to a convergence between the

Internet and traditional television sectors. According to Ovum's research, the digital TV market may be worth more than $100 billion by 2005.

TV subscription services, video gaming, and TV-based information service revenues will exceed $60 billion. Ovum expects t-commerce revenues to reach $45 billion by 2005.

1.15 Future

The future of television and digital broadband is extraordinarily bright. Rather than being eclipsed by computers or some other interface, TV, coupled with a guide that enables navigation and interactivity, will continue to delight consumers and marketers. Interactivity will lead to stronger relationships between programmers, marketers, and consumers, and the development of a user-friendly portal will ease and accelerate the pace of deployment and use. ITV plans to continue its pioneering role in helping consumers navigate the brave new digital world. T-commerce is likely to take over e-commerce very soon.

References

[1] *Compton's Interactive Encyclopedia,* Compton's New Media Inc., New York, 1992.

[2] Tichit, B., "The Thomson Experience—7 Years of Serial Digital Products and Systems," *Proc. of Seminar on Television Broadcasting and Video Production,* New Delhi, India, October 28–29, 1993.

[3] Kunzman, A. J., and A. T. Wetzel, "1394 High Performance Serial Bus: The Digital Interface for ATV." *IEEE Transaction on Consumer Electronics,* Vol. 41. No. 3, August 1995, pp. 893–900.

[4] http://www.strl.nhk.or.jp/publica/bt/en/fe0002-3.html, April 2001.

[5] Casanova, M. A., et al., "An Architecture for Hypermedia Systems Using MHEG Standard Objects Interchange," Computer Science Department, Catholic University of Rio de Janeiro.

[6] http://www.ecotec.com/sharedtetriss/news/digitalnewsitem/chapter2.htm, May 2001.

[7] "Technology Development, Hybrid Fiber coax. (HFC)," http://www.ida.gov.sg/Website/IDAContent.nsf/dd1521f1e79ecf3bc825682f0045a340/a4af3fff1af51ea3c825698-80024738c? OpenDocument, June 2001.

[8] Kramer, D., *Proc. of ITG-Sponsored Conference on HDTV* at SRG, Nov 10, 1991.

[9] Howes, K. J. P., "Satellite in the Age of Interactivity," *Via Satellite*, Vol. 11, No. 2, February 1996, pp. 114–125.

[10] Careless, J., "KA Band Satellites," *Satellite*, Vol. 11, No. 2, February 1996, pp. 62–74.

[11] "Eureka 147/DAB System Description DT/05959/III/C," December 1992.

[12] "Additional and Updated Information on Digital System for DAB, Developed by Eureka 147 Consortium and Supported by EBU," EBU Document WP 10 B/, WP 10–115/, December 1992.

[13] Schnaithmann, M., "Radio for Your Eyes: Datacasting Via DAB," *World Broadcast*, Vol. 22, No. 6, June 1999, pp. 54–56.

[14] http://www.dab-plattform.de/english/technik/dmb.htm, April 2001.

[15] Gall, D. L., "MPEG: A Video Compression Standard for Multimedia Applications," *Communications of the ACM*, Vol. 34, No. 4, April 1991, pp. 47–58.

[16] "EBU: Specifications of the Radio Data System for VHF/FM Sound Broadcasting," *Tech. 3244-E*, March 1984.

[17] Dambacher, P., "Radio Data—A New Service in VHF Sound Broadcasting," *News from Rhode & Schwarz*, 1984, p. 107.

[18] Long, G., "New Applications in Paging," *Telecom. Asia*, Vol. 7, No. 2, February 1996, pp. 35.

[19] ITU Recommendations, "ITU-R, REC. BS. 1194," Geneva, 1996.

[20] Kuroda, T., M. Yuki, and T. Tadaski, "ITU Recommendations," *Asia-Pacific Broadcasting Union: Technical Review*, No. 162, February 1996, p. 23.

[21] Kirk, D., "Broadcast Signal Formats," *International Broadcast Engineer Yearbook '96*, International Trade Publications Ltd., Surrey, England, pp. 1–6.

[22] Recommendation No. 14.3.8. "ATV System Recommendations," *IEEE Transactions on Broadcasting*, Vol. 39, No. 1, March 1993, pp. 14–18.

[23] http://www.strl.nhk.or.jp/publica/bt/en/ab0001-1.html, April 2001.

[24] Ogawa, R., et al., "Design Strategies for Scenario-Based Hypermedia: Description of Its Structure, Dynamics, and Style," *Proc. of Hypertext '92*, ACM, 1992, pp. 71–80.

[25] Ogawa, R., H. Harada, and A. Kaneko, "Scenario-Based Hypermedia: A Model and a System," A. Rizk, N. Streitz and J. Andre (eds.), *Hypertext: Concepts, Systems, and Applications*, Cambridge University Press, 1990, pp. 38–51.

[26] Hardman, L., D. C. A. Bulterman, and G. V. Rossum, "Links in Hypermedia: The Requirement for Context," *Proc. of Hypertext '93, ACM*, 1993, pp. 183–191.

[27] Brondmo, H. P., and G. Davenport, "Creating and Viewing the Elastic Charles—A Hypermedia Journal," McAleese R. and Green, C. (eds.) *Hypertext: State of the Art*, Oxford: Intellect, 1991, pp. 43–51.

[28] Liestol, G., "Aesthetic and Rhetorical Aspects of Linking Video in Hypermedia," *Proc. of Hypertext '94*, ACM, 1994, pp. 217–223.

[29] Jain, R., and K. Wakimoto, "Multiple Perspective Interactive Video," *Proc. of IEEE Conference on Multimedia Systems*, May 1995.

[30] Moezzi, A., et al., "Visual Reality: Rendition of Live Events from Multi-Perspective Videos," *Technical Report VCL-95-102*, Visual Computing Laboratory, University of California, San Diego, CA, March 1995.

[31] Jain, R., and K. Wakimoto, "Multiple Perspective Interactive Video," *Proc. of IEEE Conference on Multimedia Systems*, May 1995.

[32] Kelly, P. H., et al., "An Architecture for Multiple Perspective Interactive Video," *Technical Report VCL-95-103*, Visual Computing Laboratory, University of California, San Diego, CA, March 1995.

[33] Saied, M., et al., "Immersive Video," Visual Computing Laboratory, University of California, San Diego, 9500 Gilman Drive, Mail Code 0407, La Jolla, CA, 92093-0407.

[34] Schank, R. C., "Active Learning Through Multimedia," *IEEE Multimedia Magazine*, Spring 1994, pp. 69–78.

[35] Ming O. W. C. C., et al., "The MOS Multimedia E-Mail System," *Proc. of IEEE Multimedia*, 1994, pp. 315–324.

[36] Craighill, E., et al., "CECED: A System for Information Multimedia Collaboration," *Proc. of ACM Multimedia*, 1993, pp. 437–446.

[37] Clark, W. J., "Multipoint Multimedia Conferencing," *IEEE Communications Magazine,* May 1992, pp. 44–50.

[38] Venkat, R. P., H. M. Vin, and S. Ramanathan, "Designing an On-Demand Multimedia Service," *IEEE Communications Magazines*, July 1992, pp. 56–64.

[39] Dubery, P., and B. Titus, "Testing Compressed Signal Stream," *International Broadcast Engineer*, November 1995, pp. 32–36.

[40] CCIR Recommendation 601: "Encoding Parameters of Digital Television for Studios," *CCIR*, Vol. II, Part I, Geneva, Switzerland, 1982.

[41] Chang, Y.-H., et al., "An Open Systems Approach to Video on Demand," *IEEE Communications Magazine*, May 1994, pp. 68–80.

[42] Little, T. D. C. and D. Venkatesh, "Prospects for Interactive Video on Demand," *IEEE Trans. on Multimedia*, February 1994, pp. 14–24.

[43] Delodere, D., W. Verbiest, H. Verhille, "Interactive Video on Demand," *IEEE Communications Magazine*, May 1994, pp. 82–89.

[44] Hodge, W., et al., "Video On Demand: Architecture, Systems and Applications," *SMPTE Journal*, September 1993.

[45] Lee, M-H., et al., "Designing a Fully Interactive Video-On-Demand Server with a Novel Data Placement and Retrieval Scheme," *IEEE Transactions on Consumer Electronics,* Vol. 41, No. 3, August 1995, pp. 851–858.

[46] Fisher, S., and R. Tidrow, "Riding the Internet Highway," *New Riders Publishing,* Indianapolis, IN, 1994.

[47] Gonzalez, S., "Sound Foundations: Audio on the Web," *PC Magazine,* Vol. 15, No. 6, March 1996, pp. 209–211.

[48] Noam, E. M., "Towards the Third Revolution of Television," presented at the *Symposium on Productive Regulation in the TV Market,* Gütersloh, Germany, Dec. 1, 1996. http://www.columbia.edu/dlc/wp/citi/citnoam18.html, April 2001.

[49] Ozer, J., "Web TV Tunes In," *PC Magazine,* Vol. 15, No. 6, March 1996, pp. 129–145.

[50] Pholmann, K. C., and D. G. Lamption, "The End of the World As We Know It," *Mix,* October 1995, pp. 26–31.

[51] Richard, D., "Full Stream Ahead for Internet Broadcasting," *World Broadcast News,* June 1996, pp. 124.

[52] Craft, S., "Competition for Radio Broadcasters," *Asia-Pacific Broadcasting,* March 1996, pp. 20.

[53] Simon, D., "Music Moves Online," *Telecommunications,* Vol. 33, No. 5, May 1999, pp. 70–72.

[54] Kark, D., "Sweet Memories," *International Broadcast Engineer,* July 1995, p. 5.

[55] http://www.timbletons.com/tmc/panoramix.htm, April 2001.

[56] "Object Management Group" at http://www.omg.org, April 2001.

[57] The architecture code-named Blackbird by Motorola.

[58] Robinson, L., "A Virtual Human Agent: User Interface from Japan: Vision Meets Graphics," *Advanced Imaging,* May 1998, pp. 12–15.

[59] Shen, K., and E. J. Delp, A Fast Algorithm for Video Parsing Using MPEG Compressed Sequences, *Proc. of the IEEE International Conference on Image Processing,* October 23–26, 1995, Washington, D.C., pp. 252–255.

[60] Zhang, H. J., and W. C. Ho, "Video Sequence Parsing," *Technical Report,* Institute of Systems Science, National University of Singapore, 1995.

[61] Cover, R., "HyTime:ISO 10744:1997-Hypermedia/ Time-based Structuring Language," WG8 WWW server, as ISO/IE JTC1/SC18/WG8 N1920,1997, http://www.oasis-open.org/cover/hytime.html, April 2000.

[62] "The ACTS PANORAMA Project," http://www.tnt.uni-hannover.de/project/eu/panorama/, September 2000.

[63] Benford, S. D., et al., Crowded Collaborative Environments, *Proc. CHI'97,* Atlanta, GA: ACM Press, March 1997, pp. 59–66.

[64] Pandzic, I., et al., "MPEG-4 for Networked Collaborative Virtual Environments," *Proc. VSMM'97*, IEEE Computer Society, Geneva, Switzerland, 1997, pp. 19–25.

[65] Stewart, J., "Interactive Television at Home: Television Meets the Internet—A New Innovation Environment for Interactive Television," Research Centre for Social Sciences, University of Edinburgh, United Kingdom, August 1998.

[66] http://www.teletext.co.uk/, August 2000.

[67] "WebTV," http://www.webtv.net/, April 2001.

[68] "NetGem," http://www.netgem.com/, March 2001.

Selected Bibliography

"Additional and Updated Information on Digital System for DAB, Developed by Eureka 147 Consortium and Supported by EBU," *EBU Document WP 10 B/*, WP 10-115/, December 1992.

Andrews, J., "Audio at IBC 95," *International Broadcast Engineer,* November 1995, pp. 16–17.

Birkmaier, C. J., "A Commentary on Requirements for the Interoperation of Advanced Television with the National Information Infrastructure," http://www.eeel.nist.gov/advnii/birkmaier.html, April 2000.

Careless, J., "KA Band Satellites," *Satellite*, Vol. 11, No. 2, February 1996, pp. 62–74.

Chroust, L., "The Components of Video Servers," *International Broadcast Engineer*, September 1995, pp. 16–19.

Clark, W. J., "Multipoint Multimedia Conferencing," *IEEE Communications Magazine*, May 1992, pp. 44–50.

"Coding of Moving Pictures and Associated Audio," *International Standard, ISO/IEC 13818*, November 1994.

Craighill, E., et al., "CECED: A System for Information Multimedia Collaboration," *Proceeding of ACM Multimedia*, 1993, pp. 437–446.

Dambacher, P., "Radio Data—A New Service in VHF Sound Broadcasting," *News from Rhode & Schwarz*, 1984, p. 107.

Dambacher, P., *Digital Broadcasting*, IEE, United Kingdom 1996.

Daniel J. S., and T. A. DeFanti, "Siggraph 93 Paper—Surround-Screen Projection-Based Virtual Reality: The Design and Implementation of the CAVE," Electronic Visualization Laboratory (EVL), The University of Illinois at Chicago.

Dean, R., "BBC Sets up Virtual Reality Playpen," *World Broadcast News*, Vol. 19, No. 5, May 1996, p. 1.

Dubery, P., and B. Titus, "Testing Compressed Signal Stream," *International Broadcast Engineer*, November 1995, pp., 32–36.

Dubery, P., "Transmission in Transition," *International Broadcast Engineer,* Special 1996, pp. 26–27.

"EBU: Specifications of the Radio Data System for VHF/FM Sound Broadcasting," *Tech. 3244-E,* March 1984.

Gall, D. L., "MPEG: A Video Compression Standard for Multimedia Applications," *Communications of the ACM,* Vol. 34, No. 4, April 1991, pp. 47–58.

Goldszeft, P., and R. Rosenfeld, "The Virtual Studio System," *International Broadcast Engineer,* July 1995, pp. 22–24.

Gordon, J. H., "Digital Radio World Wide: Turning the Dream into Reality," *Technical Review; Asia-Pacific Broadcasting Union,* No.161, November–December 1995, pp. 17–25.

Howes, K. J. P., "Satellite in the Age of Interactivity," *Via Satellite,* Vol. 11, No. 2, February 1996, pp. 114–125.

"ITU Recommendations," *ITU-R, REC. BS. 1194,* Geneva, Switzerland, 1996.

Jong, G. K., et al., "HDTV Serial Interface System," *IEEE Transactions on Consumer Electronics,* Vol. 41, No. 2, May 1995, pp. 258–262.

Kenji, S., et al., "Video Disc System Using Variable Rate," *IEEE Transactions on Consumer Electronics,* Vol. 41, No. 3, August 1995, pp. 504–509.

Kirk, D., " Broadcast Signal Formats," *International Broadcast Engineer Yearbook '96,* Surrey, England: International Trade Publications Ltd., pp. 1–6.

Kramer, D., "HDTV," *Proc. of ITG- Sponsored Conference on HDTV at SRG,* November 10, 1991.

Long, G., "New Applications in Paging," *Telecom. Asia,* Vol. 7, No. 2, February 1996, pp. 35–41.

Laurent, Blonde, et al., "A Virtual Studio for Live Broadcasting: The Mona Lisa Project," *IEEE Transactions on Multimedia,* Summer 1996, pp. 18–28.

Manasco, B., "Open Architecture DBS and the Era of Customized Television," *Via Satellite,* Vol. 11, No. 6, June 1996, pp. 42–48.

Ming O. W. C. C., et al., "The MOS Multimedia E-mail System," *Proc. of IEEE Multimedia,* 1994, pp. 315–324.

"Montreux Symposia," *World Broadcast News,* Vol. 19, No. 5, May 1996, pp. 52–56.

Morta, Y., et al., " High Density Video Disc for HDTV Baseband Signal," *IEEE Transactions on Consumer Electronics,* Vol. 40, No. 3, August 1994, pp. 387–393.

Ohanian, T., "Media Composer," *International Broadcast Engineer,* May 1996, pp. 24–26.

Ostund, M., "Media Pool," *International Broadcast Engineer,* September 1995, pp. 24–26.

Plank, B., "Video Disk and Server Operation," *International Broadcast Engineer,* September 1995, pp. 30–34.

Salter, M., "Future Formats," *International Broadcast Engineer,* September 1995, p. 15.

Schank, R. C. "Active Learning Through Multimedia," *IEEE Multimedia Magazine,* Spring 1994, pp. 69–78.

Shigemi M., and A. Haruguchi, "Large Capacity D-3 Bank Video Cart," *Asia-Pacific Broadcasting Union:Technical Review*, No. 161, November–December 1995, pp. 3–8.

Toru K., M. Yuki, and T. Tadaski, "ITU Recommendations," *Asia-Pacific Broadcasting Union: Technical Review*, No. 162, February 1996, p. 23.

Venkat, R. P., H. M. Vin, and S. Ramanathan, "Designing an On-Demand Multimedia Service," *IEEE Communications Magazines*, July 1992, pp. 56–64.

"Virtual Studios Now a Virtual Reality," *Asia-Pacific Broadcasting*, Vol. 13, No. 5, May 1996, p. 29.

"Virtual TV Technology Comes to SA," *Screen*. Vol. 5, No. 2, February–March 1996, p. 15.

2

Enhanced and Interactive Television

By 2005, 28 million consumers will be reading e-books, 31 million sub-scribing to interactive television, and 37 million using digital audio.

—Andersen Consulting

2.1 Introduction

Broadcast media, such as TV and radio, are considered "passive" because the consumer passively receives the message and does not choose whether or not to view or to listen (other than by changing the channel). ITV is changing this. ITV gives consumers control over the programs they receive, as well as a bouquet of on-line services such as electronic programming guides, e-mail, e-commerce, games, interactive advertising, VOD, and Web browsing. This is achieved by creating enhanced programming and offering compelling interactive services.

Enhanced TV (ETV) broadcasting is interactive content linked to video programming by harnessing two dominant media forces together—Broadcast TV and the Internet—in a single, powerful space. In fact, enhanced TV has led to the convergence of six different media—TV, Web, CD, virtual radio, streaming video, and ITV. Any type of programming can be enhanced with icon-driven access to embedded information, usually displayed as an overlay with text and simple graphics. Enhanced content in an analog or digital signal sits waiting to be accessed by viewers.

79

ETV offers consumers a more intimate, informative, and entertaining experience. A user can make quick one-time purchases, request information on products, access additional program information, or otherwise interact with a television broadcast. This is achieved through the transmission of a URL along with the video broadcast. Companies such as Wink [1], RespondTV [2], Worldgate [3], WebTV [4], Open TV [5], Liberate [6], and CommerceTV [7] provide ETV and ITV services. When enhanced broadcast technology is linked together with other services in a unified TV portal platform, it becomes a more interesting, simple, and robust ITV application. This is especially true where enhanced broadcasts are integrated with "walled garden" environments. A walled garden is a set of Web sites and other services made available by the provider. Such an environment deals with order processing, order fulfillment, and revenue sharing issues.

The enhanced or interactive TV works in both analog and digital broadcast systems. However, bandwidth limitations place restrictions on analog ITV's current capabilities. As digital TV becomes more common and high-speed cable, satellite, and DSL technologies make their way into the home, the ITV experience is sure to change. New creative possibilities will arise. Instead of using an external network connection, digital TV is going to embed interactivity inside the broadcast signal. There is little cost in sending the interactive data out to every viewer with a digital television [8]. In the digital domain, available compression techniques allow broadcasters to send more channels by multiplexing data relating to various streams (see Section 1.4.2). As such, the addition of interactive data streams for interactivity is not very expensive, although there is much cost in setting up the system, programming it, and creating the data.

2.2 What Is ITV?

Today, ITV is interchangeably referred as Internet TV, Internet for TV, WebTV, and similar terms that invoke the familiar Internet and the Web. ITV combines the appeal and mass audience of traditional TV with the interactive nature of the World Wide Web. It allows viewing tonight's local news headlines while watching favorite network shows. It makes it possible to look up individual player and team statistics while watching a cricket show. It also allows viewers to chat with other fans or order products while watching commercials. ITV makes all this possible and is happening now!

Unlike traditional broadcast television, which sends a one-way signal into people's homes, ITV lets viewers send messages back. This makes it possible to surf the Internet on the TV screen or see specially prepared background information, such as recipes in pop-up text boxes, that has been embedded into a TV show and can be viewed by clicking on an icon with a hand-held remote control.

The new technology also makes it possible for viewers to exchange e-mail and instant messages with TV talk show hosts or other viewers, play interactive games on their television set, and buy products via the TV with a click of their remote control. Interactive television will offer viewers communications services currently provided by the Internet. One such service is the Internet relay chat (IRC) [9], which acts like a hotline where subscribers can join and talk to other people with common interests on topics ranging from religion to tips on winning video games.

2.2.1   Industry Experiences: Going Down Memory Lane

Many people believe that "interactivity" is a new and wondrous thing. However, it actually is not all that new. Broadcasters have always tried to make their shows interactive, either through postal letters, telephone lines (i.e., phone-in programs) or e-mail. Interactive television has its roots in the earliest days of television. In the 1920s, broadcasters started experimenting with interactive television. One-way video and two-way audio were test formats. In the 1950s, a simple form of interactive television was created in the CBS children's series *Winky Dink and You* (1953–1957). The interaction was created through the use of a special plastic sheet that children could purchase at local stores and then attach it to the TV set. In the program, Winky Dink, a cartoon character, went on dangerous adventures and got into lot of problems, such as being chased by a tiger to the edge of a cliff. Children were then asked to help Winky Dink escape from the tiger by drawing a bridge on the plastic screen. The show was a big hit. One obvious problem with this format, however, was that some children did not purchase the special plastic sheet and simply drew with crayons directly on the glass TV screen.

The modern era of interactive video media began in 1964, when AT&T demonstrated a picture telephone at the New York World's Fair. Over the next decade, the "photo-phone" was tested in a number of market trials and in some limited services. It was not widely accepted by consumers for several reasons, including its poor image quality, high cost, and the need for picture telephones on the receiving end. There was also little need or demand to actually see people on the receiving end in most situations. However, photo-phones did eventually find some practical applications in video conferencing and were used by many businesses and education systems.

The 1970s were rich in trials for interactive media. The National Science Foundation in the United States sponsored three major trials using interactive cable television. The areas chosen were education, community services, and worker training [10]. The U.S. Department of Health, Education, and Welfare (DHEW) supported a number of tests and services using interactive television for health care. The interactive cable TV test undertaken by Warner Amex's Qube system in Columbus, Ohio received considerable publicity. In this test, interactive data was sent in the vertical blanking interval along with the analog signals. The information contained in the data was displayed on the screen. Using the theory of transmitting data in the vertical blanking interval, the BBC developed teletext, a news and information display system.

Early experiences with the ITV trials were not very encouraging. A number of technical problems occurred. Equipment was often in a prototype stage of development and not consistently reliable. Consumer terminal equipment was generally expensive and unaffordable. Further, there were many organizational problems associated with implementing the service that had little to do with the interactive technology as such. Perhaps the trials did not adequately consider consumers' behavior and their needs and expectations. (see Chapter 6). The experiments were considered mere technological innovations. The results were obvious. Consumers did not accept the technology.

Qube was a big player in an early trial of interactive technology that obviously failed. However, it is worthwhile to examine the reasons for the failure, as well as the benefits gained by Qube in particular and society in general. The technology Qube used was very expensive in the 1970s. The cost of the consumer terminal was approximately $200, roughly four times the cost of standard decoders at that time. For the cable operations, the cable head end equipment required an additional expenditure of approximately $2 to $3 million in plant costs. In addition, it was difficult to maintain the upstream or return data path from homes, which added reliability problems for the interactive service. Production costs and interactive program design presented further obstacles. Compared with broadcast network programming, budgets for Qube programs were very low. "Interactivity" with low production values could not compete with network programming. The technology's high cost, low production value, and problems in maintaining the upstream data path were sufficient to doom the experiment.

In spite of the severe limitations mentioned above, many households subscribed to Qube's interactive service. However, how often its programming was actually viewed was generally low, albeit for a few game-format programs that drew moderate viewers and strong interactive participation. Qube's experience provided a few important lessons. The first lesson was that pay-per-view programming was potentially viable only if the costs of promoting and processing pay-per-view orders were reasonable. The second important lesson was that interactive media must be developed in a viable economic and technical context. The third lesson was that producers must learn to create with the new medium, and audiences should not be expected to change their media habits overnight.

Qube's experiments also offered many benefits. Qube worked as an important programming laboratory, introducing a number of programming formats that evolved over time and were adopted as components in cable and broadcast programming. Qube served as a tool for Warner Amex, helping to win many interactive programming franchises. In fact, the programming formats of both MTV and Nickelodeon trace their roots to Qube's experiments.

In the 1980s, TCI and Time Warner tested their own versions of interactive television in some markets. Their services enabled consumers to shop on-line, play games with people across town, and many more. Most consumers found the service very useful, but neither TCI nor Time Warner could recover the costs of operating the service while keeping prices reasonable for consumers. During the same decade, a broad scope of media offering limited forms of interaction was introduced into businesses, homes, libraries, and schools. Few had great success and the others failed completely. VCRs had extraordinary successes. Interactive videodiscs failed. Video games had moderate success. The lesson was that interactive media industries are as volatile as entertainment businesses.

2.2.1.1 ITV Trials and Services in the 1990s

A variety of players conducted interactive television trials for both cable and telephone environments. These trials involved a broad range of cable, telephone, and computer companies and offered many different strategies for delivering services. Efforts were made to provide interactive television for display on household TV sets as well as personal computers. In 1993, there were more than 500 digitally driven alliances between media (content creator), carrier (telecommunications, cable, and satellite), and computer (server and set-top box) companies. Appendix 2.A provides a list of the major trials in the United States, where television sets were used as display devices for ITV. The appendix does not include many additional trials and new services where personal computers or video-game terminals were used to provide interactive video services.

A number of telecommunications and cable TV companies launched services in the marketplace. Most provided a scaled-down form of ITV or a limited array of services. Nonetheless, these experiences were informative. It was discovered that people showed interest in movies-on-demand, interactive children's educational program, sports, and games [11]. Further, people did not increase entertainment spending for interactive services. Rather, they moved spending from one category of the household's entertainment budget to another. It was also found that while movies-on-demand were very attractive to consumers, they were willing to pay only a small increment ($1 extra) for a movie-on-demand over what they currently paid to rent a movie at a videostore. Further, consumers balked at paying hundreds of dollars for an ITV set-top box and preferred to pay a small rental fee that was part of their cable bill [12].

AT&T and Bell Atlantic conducted ITV trials in Chicago and in a Virginia suburb of Washington, D.C. The purpose was to gain experience in operating ITV networks and to get feedback about the attractiveness of such services. AT&T reported strong interest by trial homes in sports, games, and interactive educational programs for children. GTE conducted a test service in Cerritos, Calif., called *Main Street* that consisted of a database (sound and still images) of *Travel Guide, Grolier's Encyclopedia, Money Manager* software, and other content that changed little day to day. Although the project was of great research value, not many subscribed to the service [13]. TCI, AT&T, and U S West conducted a trial of movies-on-demand in Denver. The 300 test homes purchased 2.5 movies per month, much higher than the national average for pay-per-view homes. However, many homes dropped pay services such as HBO to pay for the movies-on-demand [14].

Cost Considerations

Certain early experiments around 1994 wanted to provide a huge bouquet of services (i.e., 500 channels and services ranging from electronic banking and home shopping to education and remote health care). However, the hardware and infrastructure costs to administer such interactive television services were very high, and not many consumers subscribed. The result was a handful of expensive, limited trials. Further, it indicated that the market was either not viable or would be significantly delayed. However, despite the business risks, content owners and technology and telecommunications industries are making vigorous investments, which clearly demonstrates the belief that digital video and ITV are going to provide a huge revenue stream in the future.

The cost of consumer terminals is also very important. Consumers must have the equipment at an affordable cost. A number of models have been tried. Interactive Network in California and Illinois requires a special terminal that costs a few hundred dollars and has high monthly charges. Interaction takes place not on the TV screen but on a small display attached to the terminal. Services consist of playing along with TV game shows by enhancing the original content and trying to anticipate the next play in sporting events. This model has attracted few subscribers, albeit they are reportedly quite enthusiastic. Another model, Videoway, charges a monthly subscription of under $10, no hardware costs, and interaction that takes place on the TV screen. The service offers original content, including daily interactive news programming, games, and programs for children. Videoway has developed a large subscriber base and claims to be profitable.

The most important consumer terminal is undoubtedly the microcomputer, which can serve as a terminal for many present and future interactive services. By adding sound and video through a card and CD-ROM drive, the PC can become a "multimedia" terminal. A very large share of the public is now familiar with media and machines that require interactive responses. PCs, automated teller machines, VCR remote-control keypads, microwave ovens, information kiosks at airports, and other devices in the home and workplace are teaching people important basic skills in using interactive technology. A debate has started as to whether the TV and computer can merge into a new form of consumer terminal for ITV. There are as many arguments against it as there are in favor.

Irrespective of the type of terminal, these early experiences with ITV suggest that price is of utmost importance, as users are reluctant to pay for expensive terminals.

2.2.1.2   Recent ITV Trials and Services

Although early ITV experiences were not very encouraging, they did offer good insight into the technology and what consumers prefer. Industry realized the potential of ITV, which resulted in a massive influx of capital into the interactive television sector over the last two years. A fresh rush for convergence of telecommunications, Internet, and home entertainment services to provide new kinds of ITV has taken place. Media companies, cable and satellite network operators, and ITV technology enablers have been aggressively ramping up to open new markets for interactive content and claim a stake of the expected multi-billion-dollar TV-based commerce (t-commerce) bounty.

ITV basically means enhancement of traditional TV programs. Only a very limited number of TV shows are currently encoded with interactive content, and these are gaining popularity. One such popular program is ABC's enhanced TV show *Who Wants to Be a Millionaire?* by Walt Disney. The show drew 600,000 customers in early 2000, and since March 2000, more than 9 million people have logged into the enhanced version of the popular show.

Alliance Atlantis Communications, Inc. [15], Back Alley Film Productions, CBC Television, and ExtendMedia launched North America's first interactive dramatic television series, *Drop The Beat,* at 8:30 p.m. (ET) on Feb. 7, 2000. Supported by funding from the Bell Broadcast and New Media Fund, *Drop The Beat,* a 13-part, hip-hop drama, is produced by Alliance Atlantis Communications and Back Alley Film Productions and features a companion Web site and ITV capabilities developed by ExtendMedia. Set inside the world of campus radio and hip-hop culture, this dynamic new series features "Jeff" (Mark Taylor) and "Dennis" (Merwin Mondesir), from the popular and critically acclaimed series *Straight Up. Drop the Beat* is the first dramatic series to incorporate the technology, offering fans an opportunity to click on-screen icons to purchase fashion and music items featured on the show.

Drop The Beat offers an enhanced, interactive television viewing experience through Microsoft WebTV [16]. Subscribers watching the series are able to concurrently receive on-screen links through ITV pages that appear on their screens. Viewers can then choose to connect to the *Drop The Beat* companion Web site. The accompanying Web site, www.dropthebeat.com [17], created by ExtendMedia [18], connects the creative reality of the CBC-TV drama to the external, real world. The Web site enables the show's television audience to interact directly with the show's characters through live chats, a virtual radio program, message boards, and hip-hop community building opportunities. The viewer can find out more about the characters and the actors who play them or purchase concert tickets and hip-hop paraphernalia such as the show's Universal soundtrack.

Extend Media is also producing interactive content for *Our Hero,* a teen comedy for CBC. U.S. game shows *Wheel of Fortune* and *Jeopardy,* as well as *The News Hour with Jim Lehrer,* are also designed for ITV.

AOL Time Warner has launched its own version of ITV in the United States and is planning a Canadian service that should be ready by June 2001. Vidéotron launched its ITV service to its 1.4 million subscribers in Quebec in January 2001. It offers interactive games and the opportunity to surf the Internet over the TV.

SkyDigital in the United Kingdom and Canal+ in France are driving the evolution of ITV services worldwide. BskyB has deployed 2.6 million SkyDigital set-top boxes. The ONdigital ITV service over digital terrestrial broadcasting has deployed 552,000 digital set-top boxes with telephone-line return paths. The system uses the Canal+ MediaHighway ITV system in tandem with the Canal+ MediaGuard CA system for both free and pay services. Canal+ has more than 4 million ITV subscribers among its 13 million digital satellite and cable customers across Western Europe. Table 2.1 shows the growth of set-top boxes worldwide.

In addition to these full-service trials, many companies have planned various support services and niche ITV services. For example, to help people find channels and services on large interactive systems, a number of groups have created ITV guide services like "Prevue Interactive Services," "Starsight Telecast" and "TV Guide on Screen." Another group, "Your Choice TV," has created a niche interactive service in which cable subscribers can order TV programs on demand from a library of programs, as well as from programs that were transmitted in the previous 24 hours.

2.2.2 WebTV

The 1996 release of WebTV was the start of a revolution in Web surfing. For the first time, connecting to the Internet was as easy as using a television set and an infrared keyboard. WebTV's first Internet set-top product allowed users to surf the Web on their television screens, get e-mail and read news from their sofas and armchairs. Two years later, WebTV Plus improved on this paradigm, allowing users to simultaneously view Web content and television programs, with the broadcast signal embedded in the

Table 2.1
Digital Set-Top Box Production Worldwide

Year	Number in Millions
1996	5.7
1997	7.7
1998	13.3
1999	19.0
2000	25.1

From: Dataquest.

Web content in a picture-in-picture window. Unlike future digital television (DTV), broadcasts compatible with WebTV do not carry much data in
the TV signal, but they do provide a limited preview of what DTV can do.
The broadcast signal's vertical blanking interval carries links to selected Web
pages. When a link icon shows up on the screen, the user can click on it to
summon upthose Web pages through WebTV's modem connection to the
Internet [19].

WebTV is viewable on both televisions and computers. To watch interactive TV on the television, the viewer needs to use a WebTV Plus Internet
Receiver. To watch it on a computer, the viewer needs to use Microsoft®
WebTV® for Windows® (a feature of Microsoft Windows 98) with a compatible TV tuner card. Interactive TV links work basically the same on both
receivers; however, there are a few differences in how they are handled.

2.3 The ITV Experience

Broadcasters can enrich their viewers' TV experience by creating interactive
programming that integrates Web content with traditional TV programming. Web-page content such as menus, graphics, and supporting text can
be seamlessly integrated with the TV image. Another way to get viewers
more involved with the TV show is to embed the TV display in a Web page
and trigger changes to its appearance with the TV signal. For example, during the World Cricket Series, a window on the side of the screen may show
the statistics for the current player. By sending triggers to the page, it is possible to automatically update the window to show the correct statistics every
time a new player steps up to bat. The TV audience can update the show by
interacting with the Web page. Picture this: The audience is watching a
show, a window pops up with a trivia question about the scene and, if the
viewer answers correctly, a door opens at the bottom of the screen and prizes
are offered. To create this kind of interactive TV programming, broadcasters
need to do the following:

- Design the interactive program, both the television production and
  the related Web content;

- Specify a package that contains the Web content of the ITV show;

- Broadcast interactive television links to connect the TV broadcast
  with the related Web content.

The interactive program described below illustrates how broadcasters can provide a national audience with updated, in-depth information much faster and easier than they can modify their television broadcast. While continuing to watch regular programming, viewers can access related Web pages, allowing them quick access to local information or more in-depth coverage of a specific programming.

What the Television Audience Sees

An ITV link appears as an "i" icon on the screen, depicting the availability of ITV programming for a TV show (Figure 2.1). Viewers can use their remote control or keyboard to follow the ITV link and access a new world of information. On a TV, this icon looks like the one shown below; however, on a computer, broadcasters may choose to use this icon or substitute their own.

After the viewer clicks the link (Figure 2.2), one of two things will happen:

Figure 2.1 Icon showing the availability of interactive programming.

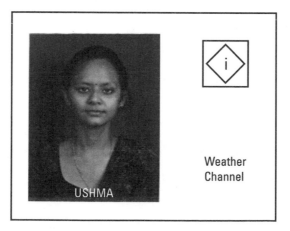

Weather
Channel

USHMA

Figure 2.2 The ITV link icon shows viewers that interactive content is available.

- On a TV with a set-top box, a selection panel appears. The viewer can choose to see the interactive program or return to the TV show.

- On a computer, interactive content for the TV show appears, skipping the selection panel step.

Both receivers can display the interactive programming specified in the ITV link, although there may be differences in the way they look. One difference is that on television, there is only one window and it is of a fixed size. On WebTV for Windows, the interactive show can open in a separate window that can be resized.

When they see the ITV link icon, viewers can select it with either their remote control or keyboard. After choosing the link, a selection panel drops down (Figure 2.3). Viewers can choose to view the interactive program or continue watching full-screen TV.

When viewers choose to see the interactive program, a connection window appears (Figure 2.4). Viewers can still watch the TV show while the interactive programming is being downloaded.

Finally, the Weather Channel program appears on the screen (Figure 2.5). Viewers can watch the current weather show live while looking at breaking weather, local weather maps, personalized weather reports, and more. The viewer can use either a keyboard or remote control to choose options from a menu that appears next to the TV display. Viewers have total control over the interactive program's look and feel. For instance, viewers can place the TV display on top of a Web page and interact with the Web site

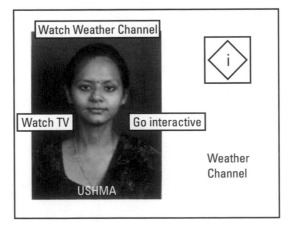

Figure 2.3 The selection panel lets viewers choose to watch the interactive program.

Figure 2.4 Connecting to the Weather Channel's interactive programming.

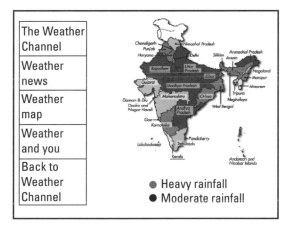

Figure 2.5 Sample from the Weather Channel's interactive programming.

while they watch the program. Broadcasters can also embed, or insert, the TV display in the Web page, reducing the display to a portion of the screen while the Web content occupies the remainder.

2.4 Why Broadcast ITV?

Widespread penetration of digital television will lead to convergence between the Internet and the traditional television sector. ITV is built on the foundation of today's television—the same video, the same commercials, and the same cable and satellite distribution. Providing users with the ability to participate actively in the programming, however, offers immense potential. ITV can also offer a great business opportunity. The following are just a few of the ways to enhance the TV viewing experience for the audience through interactive programming:

- Add depth to what's happening on the screen at any given moment.
- Add details or content that could not fit within the show's original running time, production schedule, or format.
- Reward the audience for good viewer behavior, like watching advertisements, with coupons. Provide commercial sponsors with opportunities to directly reach their audience.
- Support an e-commerce or other revenue model tied directly to the production.

Technical innovations that enable new economic opportunities and new media services are accelerating current developments in global media. This process is called *convergence*. It entails the unlimited compatibility of formerly separated technologies, services, and user habits. Interactive TV programming has much potential. However, it depends on a number of factors. One is the type of delivery mechanism. Two-way broadband cable network offers unlimited capability for return channels. Broadcasters can provide many two-way communications facilities to consumers. Digital delivery systems also help provide a bouquet of services. The compression helps provide an unlimited number of channels. In addition, the emergence of new medium opens the door to the possibility of breakthrough applications that take advantage of the medium's unique features in ways not initially apparent. In the computer world, such breakthroughs are often termed "killer apps," because they define, and indeed drive, the entire platform. E-mail and the Web browser are examples of "killer" applications that appealed to so many people and helped the Internet grow exponentially. What will those killer applications be for enhanced TV? Some speculate it might be VOD or home shopping. Other possibilities are given below.

2.4.1 Richer Viewing Experience

Enhanced TV delivers a richer viewing experience, both on television and on the Internet, to viewers. Customers can watch television as a personal TV, giving them the power to watch what they want, when they want, and the way they want. It combines an easily searchable TV listing or electronic program guide (EPG) with digital video recording (DVR). With personal TV, consumers can pause during a live TV show and return later without missing a second. Using a searchable EPG, they can search and record shows by title, type of show, even by actor. They can set reminders to record their favorite shows and then instantly retrieve those that they recorded. Other powerful features include instant replay, fast forward, and skip ahead. Personal TV is truly a revolution in television.

2.4.2 Distance Learning

Education is key to mankind's growth. Classroom education is not able to cope with the task of providing "education for all" across the globe. Distance education using broadcast media is considered an alternative to traditional schools and colleges. Classical broadcasting has the limitation of interactivity, making ITV a potential contender in the field of education. Distance learning allows the student and teacher to be in different locations. Students can experience presentations from teachers and specialists in selected fields of study from other parts of the world. Students can also interact with teachers face-to-face with the use of images and sound (multimedia), making it easier for people to learn and enjoy their classes. Students can post their problems to institutions all over the world and ask for help [20–22]. As a result, ITV can help people receive all types of lessons in their own living room, at their own pace, and in private. Several experts think that distance learning will be available on mass scale within the next 10 years.

Figure 2.6 shows an ITV program delivery. The school or college building has a cable network. TV sets with large projection systems connected to the cable network are established in each classroom. A set of video servers, video CDs, and video recorders store the teaching material. The subject teacher, with the help of the education material, delivers the course. The return channel from students allows direct interaction with the teachers.

2.4.3 Enhanced VOD

VOD allows viewers to watch any movie at any time that is convenient. The same applies to any talk show, news reports, and other programs [23, 24]. VOD also allows subscribers to request a video and have it delivered specifically to them, with VCR-like capabilities to pause, stop, rewind, and fast-forward at any time. There are also plans to sell viewers video storage space so

Educational institutions (school, college, etc.)

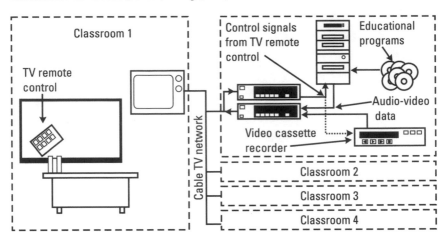

Figure 2.6   ITV for education.

they can save their home movies on a central video server and play them at their leisure. Many companies currently provide VOD using a system similar to ordering pay-per-view movies by telephone and receiving them on a special cable channel.

When cable services switch to digital transmission and are able to use enhanced data compression methods, VOD systems are expected to greatly improve [25]. In addition, subscribers can choose from about 500 channels, including channels that offer movies, sports, shopping, cooking, and children's programs (known as specialty channels) that show the same programs to all viewers.

2.4.4   Information Retrieval

ITV facilitates subscribers to get information on everything they want. Subscribers can get phone numbers, travel guides, airline schedules, product information, and even coupons for supermarkets by using a printer linked to the ITV system. Another possibility is to browse the Internet. Subscribers can receive local and international news, traffic reports, weather, sports scores, and stock quotes every morning either from the TV or printed on an optional letter-size printer. Personalized news tailored to the viewer's choice can be delivered. Further, the service allows quick tracking of stocks and sports teams of interest by providing customized details on both [26].

2.4.5 Game Shows

ITV allows viewers to participate in game shows. The TV host can ask a question and the viewer can answer using a keyboard that is connected to the ITV system. Game shows can run concurrently with live cricket games, and the viewers may be asked to predict the next outcome. Viewers can compete to be the top scorer and send their scores for a national ranking. In addition to sports games, game shows are a popular venue for ITV. Viewers can even win prizes by getting a certain number of points.

2.4.6 Audience Targeting

 The power of the Internet, coupled with ITV, enables broadcasters to target specific audience segments that it wants to reach with far greater accuracy than traditional broadcast TV.

2.4.7 Audience Feedback

With a two-way broadband connection, audiences can participate in shows, chat with other fans and with their favorite stars, and purchase related products. Broadcasters can gain immediate feedback about the audience, giving broadcasters and sponsors more ways than ever to build and retain relationships.

2.4.8 New Revenue Models

New revenue streams can be tapped with ITV through sponsorships and advertising. ITV is a direct link to customers for merchandising. Memorabilia, movies, games, toys, clothing, and much more can be offered directly to the viewer through ITV.

2.4.8.1 T-Commerce

People have been talking of m-commerce and e-commerce. The latest buzzword making the rounds of the digital set is something called t-commerce, shorthand for television-based commerce. ETV has the potential to evolve into the platform for the next-generation of on-line viewers. TV audiences can buy products and services with the click of a remote control. There are currently 170 million PCs in the world, almost 500 million mobile phones, and almost 1.5 billion television sets. Imagine combining 1.5 billion TV's, which have a proven ability to sell, with the Internet's power to conduct sales instantaneously. It gives new meaning to the phrase "impulse buying."

The following are salient features of t-commerce:

- It is a new way for consumers to purchase the products they want, capitalizing on the universality of television and the familiarity of the remote control.
- It is based on the mindset and behavior of the consumer. T-commerce is so easy, it's an impulse purchase (i.e., "That looks cool. I think I'll buy it.") Without changing the channel, lifting the phone or turning on the computer, a purchase is made.
- It is ready for the mass market. Completely safe, secure, and private, it can handle the tremendous waves of consumer transactions.
- It is free for subscribers, as far as this additional benefit of shopping is concerned. Unlike e-commerce, subscribers don't have to pay for Internet uses and line charges.
- It is ideal for cable companies that are moving aggressively to install digital set-top boxes in consumers' homes.
- It is easy. Viewers see an icon on the screen whenever commerce opportunities are offered. Using their remote control, viewers can then make purchases, "save" items for later consideration, request more information such as discount coupons, and even check the status of previous purchases. A host of more advanced capabilities also exist.
- It is built for high capacity and can handle sharp traffic bursts.
- It is easily scalable, from small deployments to large.
- It is compatible with digital cable and digital satellite broadcasting systems.

Furthermore, t-commerce does the following:

- Allows sophisticated consumer profiling and database development for pinpoint targeting of special offers;
- Offers a complete transaction, everything from choosing a color and size to charging the item to a credit card in a few simple steps;
- Links directly to partners and suppliers for seamless integration;
- Allows programmers and advertisers to decide which offers they want to make during which shows.

2.4.8.2 Shopping Via Database Search

T-commerce could offer a multimedia shopping magazine service that describes the products offered by a book and record shop. The video material could be transmitted in the 1/4-size format to reduce the bit-rate requirements. The multimedia presentation could be accompanied by text—updated throughout the program—giving further information on the product being discussed or on special offers. By selecting an on-screen button, a viewer can put the selected item in a virtual shopping basket. At any time, the customer can return to the basket and scroll through a list of the contents using the up and down arrows on his remote control, ask for details of any tagged product, or initiate a purchase sequence (once again, requiring financial authority).

A similar selection process could be used to set up the parameters for a search, perhaps to find a gift item. In this case, the search parameters would be compared with available products on a remote database, and a list of potential gift items would be delivered to the set-top box for display. Detailed information could be delivered in still picture, audio, or textual format to the box on demand. This is strictly an on-line activity, but retaining a broadcast background is found to be desirable. This can be done using the magazine service that originally attracted the customer's interest or (in the case of satellite, where there is sufficient capacity) using specially transmitted general-purpose background audio or video material.

2.4.8.3 Detailed Product Information

Interactivity can be used to deliver greater detail about products and services. When a television program or advertisement is running, viewers may want more information. ITV content gives broadcasters and advertisers an avenue to reach viewers. Instead of pointing viewers to a Web site, the Web site is melded into the television experience.

2.4.8.4 New Form of Commercials

New forms of commercials are being tried in ITV. The traditional 30- to 60-second TV spot still exists but with the addition of an interactive icon. What happens when the viewer clicks on an icon is up to the advertiser. Results can vary from the simple fulfillment of a request for a free sample to taking the viewer away from the linear stream to an information micro site. Every subscriber to these interactive platforms has an ID number that is unique to that individual. This allows the platform operator and the advertiser to track every move an individual makes while using the service.

Britain's Royal Air Force (RAF) is running a recruitment ad featuring a site that allows viewers to play a game of refueling and rearming a fighter. If viewers can refuel a plane in 30 seconds, they can apply, on-line, for an interview with the RAF [27].

2.4.8.5 Horse Racing

A customer can select the interactive domain while viewing a horse race in order to place a bet. The broadcast TV picture is reduced to 1/4 its size to provide screen space for further information while retaining contact with the original service. The customer then requests information on a particular horse on which he wants to bet. A still picture of the horse with accompanying text is delivered to the customer. If the customer wants to browse through the text stored in the set-top box, a complete screen with broadcast audio is provided. Alternatively, the customer can request a verbal description of the horse and its recent performance. When the customer has completed his selection, he can proceed to place a bet subject to financial authority, most probably achieved through the entry of a personal identification number.

2.4.9 Voting

Subscribers can participate in local, regional, or even national polls simply by pressing a button on the remote control. Subscribers can be given an individual password, and encrypted voting can be sent to a central place. The results can be quickly tabulated and shown on the screen.

2.4.10 Telebanking

Until now, payments were made using traditional payment mechanisms like regular bank accounts and credit cards. Now consumers of ITV can open a special electronic cash (e-cash) account to pay for services. In this system, monthly fees for a service is transferred from the user's bank account to the merchant's account. The account can be used to pay for items purchased from home shopping. Another way is to pay by credit card. In this case, credit card details are sent to the service provider involved and the credit card organization handles the payment. Again encrypting, scrambling, or coding confidential information must be used [28].

2.4.11 Walled Garden

The walled garden is another buzzword for the ITV environment. The walled concept cuts two ways. First, it keeps subscribers in, directing them to content providers that are either paying or splitting revenue with the platform operator to be featured inside the garden. At the same time, the wall keeps the rest of the Web out, limiting selection to the paying merchant. In the walled garden, people can buy movie tickets, CDs, DVDs, do their banking, or conduct all kinds of e-commerce transactions. The content within the walled garden is designed for the TV and is easy to use. In contrast, material on the Web just does not look as good when viewed on a TV screen. Forrester Research predicts walled garden commerce will reach $12.5 billion by 2005.

2.5 Different Types of ITV

ITV services provide a variety of facilities, such as personalized channels, t-commerce, telebanking, etc. This has led to a number of nomenclatures for ITV according to the services.

Enhanced TV

Any type of programming can be enhanced with icon-driven access to embedded information, usually displayed as an overlay with text and simple graphics. These embedded text and graphics can be accessed as a full screen or "page" of text and graphics. Enhanced content in an analog or digital signal sits waiting to be accessed by viewers, so a return path is not needed. Key vendors in this space include Wink, Worldgate, WebTV, OpenTV, and Liberate.

Internet TV

All the functionality of the Internet, especially Web browsing and e-mail, is delivered to a TV screen. The visual quality of text is poorer on analog TV compared with digital TV. If a telephone line is used for the return path, Internet access is temporarily suspended for incoming calls. Key players in this market are WebTV, Worldgate, and Liberate.

On-Demand TV

Any kind of programming can be offered on demand, from movies to weather news to stock markets. A video file server plays back content on request within a digital two-way system. Ideal for pay-per-view services, content can be seen whenever viewers wish. Users are no longer passive watchers on "appointment TV" with schedules fixed by TV stations. To date, there are serious technical limitations in terms of bandwidth and transmission capabilities in providing such service. However, with networks such as broadband two-way hybrid fiber coax (HFC) cables or DTH with return channels in Ka-band in place (see Sections 1.6 and 6.4), such services will be common. Many see VOD as the "killer application" for interactive TV. Major vendors include Concurrent, DIVA, Oracle/Liberate, and Seachange.

Individualized TV

Individualized TV allows subscribers to modify a program to match their individual desires. The control allows changing camera angles at will, calling up instant replays in sports and live news, guiding the plot in dramas and comedies, and having the TV host respond to answers on a game show. Individualized TV content was first developed for one-way broadcasting but can now be two-way. The technology first came from ACTV, but now OpenTV is also providing individualized TV.

The development of devices that enable people to customize TV viewing to their interest could fundamentally change the way the TV industry has operated over the last 50 years. Forrester expects viewership of TV advertisements to drop nearly 50% over the next decade, forcing advertisers to accommodate to this new environment. This could result in recognized brand names moving away from traditional advertising to increasingly sponsoring thematic channels, including learning channels. They could also move to an information mode of advertising, where it becomes critical to be "captured" during key word searching through personal video recorders (PVR). This mode would not be dissimilar to Web-based advertisements.

Personal TV

Personal TV is the term coined for the PVR that records programming by title, timeslot, rating, actors, or theme on a hard disk. PVRs use an intelligent interface and an internal hard drive to digitally record programming in anticipation of viewer demand. They could also be capable of holding 100 hours or more of programs. With full VCR functionality, the personal TV can pause during a broadcast as content is cached on the disk. It also has the ability to skip over commercials. If the personal TV has a return path, it can support pay-per-view billing. With features like instantaneous fast-forward and reverse, the ability to pause live broadcasts, and easy-to-create viewer profiles, PVRs will broaden viewer options by offering, at any given moment, a menu of recorded programs based on the viewer's preferences. Leading vendors are TiVo and Replay, Echostar, WebTV, and Pace.

The personal TV has three stages of evolution:

- A smart videocassette recorder that can seek out and record regular broadcasts, like TiVo and Replay TV do today.

- Virtual channels, where networks create entertainment packages that can be accessed on demand. For example, a broadcaster might decide to create a channel devoted to action movies. If subscribers wanted that action pack, it would then be automatically downloaded to the disk in their ITV receiver. The channel creator could also add other information with the download, like Internet links or related merchandise offerings.

- By about 2005, the disk drive will hold as much as one terabyte or a thousand gigabytes. That means consumers will have the ability to record a thousand hours of digital video. They can then choose the time and date of viewing.

Play TV

PlayTV encompasses interactive video games on television involving single and multiplayer competitions. The Sega channel is one example, where games are downloaded over cable to the Sega player device at home and played interactively. Another example is the NTN, Inc. system. PlayTV is expected to be very popular within 3 to 5 years.

Banking and Retail TV

Electronic banking and e-commerce applications on the Internet are being ported to the television. Interactive advertising will allow viewers to request e-mail brochures and actually order products on screen. The broad range of shopping services may eventually eclipse VOD as the killer application. The leading channels are QVC and the Home Shopping Channel.

Educational TV

All forms of ITV can be applied to educational services for all types of formal and nonformal education. "Education for all" is the motto of nations worldwide. As such, distance learning will be a key issue for policy makers. "Cable in the Classroom" and "the Educational Satellite Consortium" are major players.

Community TV

ITV lends itself to local community involvement, everything from town council meetings to electronic voting. Just as the Internet is host to evolving "virtual communities" of shared interest among scattered people, ITV, with the addition of two-way video telephony, will result in the sharing of ideas, thoughts, and interests among various groups.

Global TV

As the network builds out globally, there will be increasing demand for access to international programming with automated language translations. ITV will be a mass medium for local-to-global cultural exchange.

2.6   Transmissions and Range of Services

The selection and combination of services to offer is the most difficult decision for broadcasters. How can they transmit their services (down transmissions), and how can users access them (up transmissions)? Surprisingly, there are not many choices.

Down Transmissions

1. Vertical blanking intervals (VBI) in conventional analog TV broadcasts (64 Kbps);
2. Data channels offered by analog satellite broadcasts and others (224–1120 Kbps);
3. Digital terrestrial and satellite broadcasts;
4. Digital cable television;
5. Telephone lines.

Up Transmissions

1. Reception server connected to existing phone lines (including ISDN);
2. Internet (via conventional phone lines, ISDN, cable modem). Many cable modems have 10-Mbps or 30-Mbps transmission capacity);
3. Interactive-type cable television.

Video transmissions via the Internet using telephone lines cannot be compared with conventional television in regard to display size, picture quality, and service contents. For example, down transmission services by telephone lines (28.8 Kbps) and ISDN (64 Kbps) beyond text, still pictures, and audio services are limited in terms of fidelity, speed, and capacity for moving images. Even at the level of high-speed ISDN transmissions (1.5 Mbps maximum), such transmissions are a long way from matching the capacity and picture quality offered by existing television broadcasts (20–50 Mbps).

Down transmission by digital satellite rather than transmission by wire has been drawing great interest. Previously, hard-wired systems were the preferred method of transmission, but radio transmission systems have been reevaluated and deemed superior due to their practical and economic advantages.

Wired Network

In interactive broadcasting, the basic image of software production can be symbolized by the Web plus network. This may result in the evolution of wired networks, provided by newspapers, telecommunications companies, and broadcasters, for the general public. A wired network will be a two-way pipe to every home, such as HFC or a digital transmissions system with an in-built return link. They would rely on databases and on-line service use. However, such systems have not yet evolved from terrestrial and satellite broadcasts, cable TV, or digital satellite broadcasting. As changes take place in multimedia and media maintenance and it can be provided at lower costs, conventional communications companies are likely to expand services to the general public from those information services previously available only to fellow businesses. Obviously, there is a high potential to form wired networks, which could be used to cover extensive areas, ranging from hobbies and entertainment to news. The coming competition will determine which transmission systems the wired network media will use to serve up and down transmissions. Enhanced DTV will offer a number of ways to do that:

- Synchronous enhancements that are transmitted for use at specific times in a program;

- Always-on enhancements, such as navigation bars, that are constantly accessible at the click of a remote control or mouse;

- Asynchronous or post-broadcast enhancements that are silently transmitted into the DTV receiver's memory and can be activated when the viewer chooses;

- Interpolated enhancements that the viewer can choose to insert seamlessly into an ongoing program.

Unfortunately, any extensive enhancements that would need to be stored in the receiver's memory will have to wait until DTV sets commonly include a hard disc, like the TiVo and Replay VCR-substitutes now on the market.

2.7   Who Can Watch ITV?

ITV can be received by using a set-top box, the remote control, and the television. The subscriber also needs to be connected to a network operator providing the services.

2.7.1   Set-Top Box

For all delivery mechanisms, a set-top box has to be connected between a normal TV and the incoming signal. This decodes the signals as they arrive from the main control center. The set-top box serves a number of purposes:

- It converts the digital signal into a format that can be viewed on the TV.

- It usually provides the means by which only registered users can access the digital broadcasts. This is known as conditional access.

- It contains the software to allow the use of interactive services.

- It may have an internal modem that can be connected to a telephone line, both for the return path for interactive services and to send user statistics back to the service operator.

- It currently has limited memory to hold updated versions of the electronic program guide and other data that might be downloaded into the box. Later generations of set-top boxes are likely to have considerably more memory.

There is no standard specification for a set-top box as this is determined by the service broadcaster, who may sell, rent, or give away the set-top box when a household subscribes to a particular service. There are many different types of set-top boxes available. They vary according to two particular features:

- Conditional access system;
- Application program interface (API) software.

2.7.1.1 Conditional Access

Conditional access is the method by which the service broadcaster can control access to the bouquet of channels that are subscribed to by the viewer. A viewer subscribing to a service is usually provided with a smart card that is inserted into the set-top box. It is also a useful way of controlling which regions are allowed access to particular programs, allowing fees for intellectual property rights to be more finely tuned. If the consumer decides to subscribe to a different delivery system because he wants additional channel choices, he may very well need a different type of set-top box.

When more than one bouquet of offerings is available via the same platform, for example, satellite, simulcrypt agreements are usually reached, allowing set-top boxes with different methods of conditional access to receive both bouquets of offerings. Simulcrypt allows different decoders with different conditional access methods to decode different bouquets of channels from different service providers, if the appropriate smart card is present. Another method is multicrypt transmissions that allow two different encryption systems to coexist in the same receiver. The signal is sent sequentially through different conditional access modules that are in the set-top box. If the appropriate smart card is present, the viewer will be able to receive that bouquet of programs. However, set-top boxes capable of using a number of different conditional access systems tend to be more expensive.

2.7.1.2 API

The API is essentially the software operating system for the set-top box. In particular, it controls the types of interactivity that are possible. There are currently a variety of proprietary APIs in use. It is likely that the differing APIs will migrate toward interoperability in line with the path set out by the DVB Group [29] Multimedia Home Platform (MHP) standard, which appears to be moving toward a Java-based standard (Section 3.6). This will include Internet standards, making different APIs compatible with each other and the Internet. Software interoperability will lead to increased development of interactive services on digital television platforms.

2.7.2 Remote Control

Users need an interface to find their way through all the services available to them and to communicate with the main control system. This can be done by a menu system (on-screen graphics). This allows users to select a service and then specify what particular aspect of that service they require. The user has to learn how to use such interface and all its functions.

2.7.3 Television

In order to get the ITV system at home, users can either use their existing SDTV or HDTV, which has the benefit of providing a higher resolution picture as well as CD-quality sound [30, 31].

The above is just the basic equipment needed to get services like home shopping, EPGs, and information retrieval. For services such as printing out coupons, users will need a printer in addition to the set-top box, TV, and remote control. While many companies are racing to join the digital TV market, the technology's potential is too new for standards to be in place for such elements as conditional access and EPG systems. This presents a dilemma for a manufacturer or designer that wants to be in the market but is wary of starting out with what will become proprietary technologies or processes [32].

2.7.3.1 TV or PC

ITV is similar to PCs in certain aspects. Both offer interactive services. ITV offers home shopping, distance learning, and Internet access. PCs have on-line shopping and Internet access, too. Of course, PCs can offer hundreds of channels like ITV simply by adding an appropriate tuner card. PCs also have processors that are more advanced than those in ITV set-top boxes, allowing them to create and run programs such as word processors, databases, spreadsheets, and other applications. ITV does not have such a feature yet. It is, however, cheaper than the PC. One very important aspect of TV is that it is watched from a greater distance. TVs have large screens for group viewing from across the room, and PCs (usually) have smaller screens for up close, one-on-one viewing . With a large-screen monitor on a PC, users can view it from across the room. This suits viewing TV-style programs, but it probably would not be good for one-on-one PC-style uses, since it uses up the entire room for just one person's activity. So the choice of a PC or TV depends on the type of application used and the user's preference. We will discuss this topic further in Section 3.3.6.6. One thing, however, is certain. ITV will not merge with PCs. Regarding this connection, it is worthwhile to quote Bill Gates and Jerry McCarthy.

> *...the form and function of communications appliances such as TVs, computers, and telephones will tend to remain distinct - because we use the tools in different ways.*

—Bill Gates [33]

> *Television sets and computers will never become one unit. Sure, there will be more interactivity in television sets......But we look at interactivity that complements TV viewing rather than supplants it.*

—Jerry McCarthy, president of Zenith Sales Company [34]

2.7.3.2 Integrated TV Sets

Integrated TV sets incorporate the set-top box. An integrated TV is also a HDTV in a 16:9 format, making it possible to view pictures in a wider format that is more natural for the eye, when that format is broadcast. However, not all HDTV sets sold in Europe have integrated set-top boxes, because the 16:9 format has been broadcast in some countries over an analog signal. However, in the United States, HDTV sets were not introduced until the end of 1998 and do have integrated set-top boxes. Some units in the United States have a separated HDTV system with a receiver-decoder unit and a display unit. A number of manufacturers are now offering or preparing to launch integrated digital TV (IDTV) in the United States or European markets, including Mitsubishi, Philips, Panasonic, and Sony.

2.8 Key Players in ITV Service

The key players in ITV broadcasting are:

- *Content providers:* individuals or organizations with a specific interest in the production and management of content;
- *Distributors:* organizations that hold a license to broadcast, are legally liable for the content, have a primary relationship with the

consumer, and, generally, collect advertising revenues and subscription fees;

- *Carriers:* organizations that provide the delivery system to broadcasters;

- *Hardware suppliers:* companies that supply equipment to the consumer;

- *Software suppliers:* companies that supply the API software in the consumers' set-top box that enables interactive services to be deployed;

- *Interactive service software developers:* companies that develop the interactive service software based on the API used in the set-top box;

- *Management solution providers:* organizations that enable content aggregation, distribution, and presentation using interrelated hardware and software solutions for total management.

Some players may take on a number of these roles.

2.9   Future of ITV

The future of television and digital broadband is extraordinarily bright. Rather than being eclipsed by computers or some other interface, TV, coupled with a guide that enables navigation and interactivity, will continue to delight consumers and marketers. Interactivity will lead to stronger relationships between programmers, marketers, and consumers, and the development of a user-friendly portal into this world will ease and accelerate the pace of deployment and use.

With the advent of newer technologies, the future television may not be what it used to be. Instead of a flat-screen version, consumers may have a 3D television, where they can view a program at any angle, at a size they prefer. ITV will mean more than just being able to see and hear. Through sensory equipment, consumers may be able to touch, smell, and even taste through ITV.

Virtual reality can play a vital role in the future of ITV. Consumers can participate in virtual reality games. In education, students can participate in active learning through virtual worlds. Consumers can take part in historic battles or find out how blood cells travel in the human body.

We are still at an "experimental" stage, where all types of systems and uses are being tried out. The most important step for the future is to move to real markets with standards and infrastructure in place. Many services are also becoming available via the Web. In fact, interactivity through the TV and through the Web are bumping alongside each other. However, a key point needs to be made when comparing Web-based developments with those of interactive TV. The Web has evolved with end users having a lot of control over the content and what they want to select. They can also easily publish material themselves. Interactive services via the TV will develop in a more traditional manner, just like TV programming, where end users have not been involved in the innovation process and have not been able to express their interest in interactive services or systems through purchasing and use in the home. The impact of this on new interactive services has yet to be realized and will be a key factor on whether they will be able to compete with Web-based services. We will discuss the future of ITV in detail in Chapter 6.

2.10 Conclusion

ITV is going to change the way we watch television. ITV offers a host of services, from selecting a movie to personalized news to e-mail to Internet relay chat. These services are interactive, allowing the viewer to communicate, not only by receiving signals but by the ability to send signals as well. This two-way communication allows viewers to access services like VOD, distance learning, and home shopping. Internet access makes all the Web's services and facilities available from the TV.

Communication through ITV can help consumers reach other people around the world. This is particularly useful for people who are handicapped. Through ITV, these people can interact with other people without leaving their beds. They can make friends and search for information at the click of a button. They can buy their groceries through home shopping. In the past, these people were isolated from the world, unable to make any contact except through the window or an occasional walk out of their homes. Through ITV, they can be part of the community again.

Currently, the average person spends about 20 hours a week watching television [35]. Through ITV, people can do many things from their own homes, such as home shopping and distance learning. This, together with the availability of more TV channels and VOD, is going to make the number of hours people spend in front of the television much longer. The impact of this will need watching. However, ITV has immense potential, and it is for the content providers and program developers to explore the possibility of this new brave world [36].

References

[1] http://www.wink.com, April 2001.

[2] http://respondTV.com, April 2001.

[3] http://wgate.com/splash/main2.html, April 2001.

[4] http://www.webtv.com, May 2001.

[5] http://www.opentv.com/interactive/, April 2001.

[6] http://products.liberate.com, April 2001.

[7] http://www.commerce-tv.com/home_sr_flash.html, May 2001.

[8] http://www.pbs.org/opb/crashcourse/enhanced_tv/experiments.html, April 2001.

[9] http://www.mirc.co.uk/irc.html, March 2001.

[10] "Symposium on Experiments in Interactive Cable TV," *Journal of Communication*, Spring 1978.

[11] *Wall Street Journal*, July 28, 1993, p. B1.

[12] *Wall Street Journal*, Oct. 14, 1993, p. B1.

[13] *Los Angeles Times*, Aug. 31, 1993, p. A1.

[14] *Wall Street Journal*, Oct. 14, 1993, p. B1.

[15] "Alliance Atlantis Communications Inc.," http://www.allianceatlantis.com, April 2001.

[16] http://www.microsoft.com/tv/contentdev/building.asp, April 2001.

[17] http://www.dropthebeat.com, April 2001.

[18] "ExtendMedia," http://www.extend.com, May 2001.

[19] "Web TV Plus," http://www.current.org/tech/tech903.html, April 2001.

[20] "Social Aspects of the Web," http://www.cs.unc.edu/wwwc/public/dilip/socialAspects.htm, April 2001.

[21] "Distance Learning," http://www.brightok.net/dl.html, April 2001.

[22] "Distance Education and the WWW," http://tecfa.unige.ch/edu-ws94/contrib/peraya .fm.html, May 2001.

[23] "The Richter Scale—Interactive Television," http://www.strokeofcolor.com/richter/ pcgr/pc931116.htm, March 2001.

[24] "Video On Demand," http://www.ee.ed.ac.uk/~me94te/vod/vod_index.html, March 2001.

[25] "Zen and the Art of Interactive TV," http://www.cs.ubc.ca/spider/hoppe/zen.itv.html, April 2001.

[26] "Interactive Systems, Inc.," http://www.teleport.com/~isi/, April 2001.

[27] Jenkins, Bob, "Interactive Clicks and Games in the U.K.," *Multichannel Internationals,* November 2000, p. 10.

[28] "Working of Telebanking," http://wwwedu.cs.utwente.nl/~aitnl002/project/teleb.htm, April 2001.

[29] "DVB," http://www.dvb.org/, April 2001.

[30] "Public TV and the Transition to Digital HDTV," http://www.current.org/atv1.html, May 2001.

[31] "HDTV EUROPE," http://www.microresearch.be/hdtv/index.hts, December 2000.

[32] http://www.ntl.co.uk/guides/digitaltv/interactive.htm, April 2001.

[33] "Bill Gates and the Internet PC," http://igubu.saix.net/speed/intpc.htm, April 2001.

[34] "TV or PC?", http://www.appliance.com/app/tv_or_pc.htm, March 2001.

[35] "IPA Trends in Television," http://www.ipa.co.uk/reports/1996/1/table1.htm, April 2001.

[36] "Social Impacts of Information on Demand," http://www.contrib.andrew.cmu.edu/~radar/telecom/information/social.html, April 2001.

Selected bibliography

"Broadcast Spectrum and the Debate on the Future of Television," http://cdinet.com/Benton/Retrieve/bspectrum.txt, May 2001.

"BT News Releases," http://www.bt.net/home/newsroom/document/nr9535.htm, April 2001.

Cronin, G., "Marketability and Social Implications of Interactive TV and the Information Superhighway," *IEEE Transactions on Professional Communications,* Vol. 38, Issue 1, March 1995, pp. 24–32.

"Far West Laboratory for Educational Research and Development," http://www.fwl.org/edtech/ distance.html, April 2001.

http://www.doc.ic.ac.uk/~nd/surprise_96/journal/vol4/khp1/report.html, April 2001.

http://www.microsoft.com/ISN/Cable/, April 2001.

http://www.microsoft.com/tv/content/, March 2001.

"MPEG: Standards, Technology, and Applications," http://www.dse.doc.ic.ac.uk/~nd/surprise_96/ journal/vol2/sab/article2.html, April 2001.

"Opening the Set-Top Box Market," http://farnsworth.mit.edu:80/Pubs/settop_mkt/abstract.html, March 2001.

"Real Time Interactive—On-Line Television." http://www.real-time.co.nz/tv.html, April 2001.

"The Future of Retailing/Home Shopping," http://www.duke.edu/~bjones/lit/outline.html, April 2001.

Appendix 2.A

Partners in ITV Trials (1994–1996)

Table 2.A.1
Cable Trials

Company	Location
AT&T and Viacom	Castro Valley, CA
IBM, Videotron, and Hearst	Montréal, Quebec, Can.
Rogers Cablevision and Microsoft	Ottawa, Ontario, Can.
SW Bell and Cox Enterprises	Omaha, NE
TCI and Microsoft	Denver, CO and Seattle, WA
Time Warner	Orlando, FL

Table 2.A.2
Video Dial-Tone Trials (1994–1996)

Companies	Locations
Ameritech	Chicago, IL; Cleveland, Columbus, OH; Detroit, MI; Indianapolis, MD; Milwaukee, WI
Bell Atlantic	Arlington, VA; Dover, Florham Park, NJ; Washington, DC
Bell South	Brentwood, TN; Heathrow, Hunter's Creek, FL; Chapel Hill, NC
GTE	Clearwater, FL; Honolulu, HI; Manassas, VA; Thousand Oaks, CA
NYNEX	New York, NY; Portland, ME; Warwick, RI
Pacific Bell	Los Angeles, Orange County, San Diego, San Francisco, CA
Rochester Telephone	Rochester, NY
SNET	West Hartford, CT
SW Bell	Richardson, TX
U S West	Boise, ID; Denver, CO; Minneapolis, MN; Portland, OR; Omaha, NE

3

Technology and Standards

3.1 Introduction

Terrestrial analog television broadcast in the VHF or UHF bands is still the major method of delivering video entertainment to the home. The signal is received with a conventional analog television receiver using one of three color television standards, PAL, NTSC, or SECAM [1]. Conventional television uses analog transmission, which means the signal level transmitted varies in accordance with the color and brightness of the pictures' elements. Any disturbance to this signal along the transmission path introduces false information, or removes information, that cannot be repaired, reducing received pictures to lower quality than those transmitted.

In the early 1980s, the ITU started developing global standards for the new generation of television broadcasting. One such specification was developed to provide a viewing experience similar to that offered by wide aspect ratio (16:9) pictures with at least double the resolution of analog television services [2]. The service was intended to provide the viewing experience offered by 35-mm film in the cinema. Throughout most of the 1980s, the initial focus of this work was on increasing the capability of analog technology.

During the past 35 years, there have been major developments in digital techniques to transmit, store and process images, sound, and data. Digital signals are converted to analog for transmission. Much of this progress has been made possible by advances in computing technology and microcircuit design. Most television studios have used digital technology for the last five years or so. Advances in the compression of digital video have allowed its transmission, and many countries have started regular digital broadcasts. In digital transmission, errors, within certain limits, can be identified and corrected, thereby eliminating interference. The received picture is a faithful replica of the transmitted picture. The major obstacle with digital systems has been the difficulty in processing and transmitting the large volume of digital data needed to faithfully code a moving picture.

Traditionally, television transmissions had been via "over-the air" broadcasts. In mid-1980s, the continual expansion of the telecommunications business encouraged research into satellite, cable, and optical fiber as a method of carrying communications signals. These mediums promised an enormous increase in capacity but needed digital technology to reach their full potential [3]. Internet broadcasting has also recently caught the attention of broadcasters.

The computer industry was also expanding rapidly as a result of improvements in materials technology that allowed larger integrated circuits and chips for processing and electronic memory to be manufactured with increasing levels of capability at lower cost. Both the computer and telecommunications industries were heading toward capacities large enough to handle video images routinely. By the end of the 1980s, the computer industry was moving ahead with faster and larger data handling capacity, and digital computers were increasingly penetrating the mass consumer market. Common carriers were already using optical fiber for many applications and could see the delivery of television as a major growth area that could generate considerable new business for their transmissions capacity.

Research by industries had shown that digital compression of video, and new digital modulation schemes for satellite, cable, and terrestrial services, would facilitate high-quality multichannel reception. These technologies promised to offer improved efficiencies and reliability. The turning point in shaping the future of the television business came at the beginning of the 1990s, when the U.S. Federal Communications Commission (FCC) decided to encourage HDTV in existing broadcasting bands. In November 1991, the ITU established a special task group (TG11/3) to pull together a world standard for digital terrestrial television. The subsequent work of the American Grand Alliance helped develop a special composite system [Advanced Television Systems Committee (ATSC)] in digital rather than analog form [4].

In Europe, the HD-MAC [5] modulation program was largely abandoned in favor of the DVB project that quickly developed a set of digital television standards covering all main forms of delivery—satellite, cable, and terrestrial [6]. The hunt for HDTV had irrevocably turned from analog to digital. The changeover to digital television first began in the subsector of satellite television, being driven by the new satellite pay-TV services and their recognition of the improvements in efficiencies (lower transponder costs). Digital cable television applications followed to allow transparency between satellite and cable, again in the pay-TV domain. These distribution and delivery services used digital technology and DVB standards for its bandwidth efficiency and its flexibility.

In the same period, the introduction of digital technology to the production and reception parts of television broadcasting dramatically changed the traditional broadcasting model, moving broadcasting into the mainstream of telecommunications and computing and enabling major innovations in services available to the viewer. In production, the traditional linear approach for editing and program assembly started being replaced by the nonlinear, random access approach enabled by large servers, compression technology, and sophisticated high-speed broadcasting and telecommunications networks. The monolithic production facility could readily be distributed into specialized service units linked by the network over large geographical areas. In production techniques, a number of experiments concerning immersive video [7], hypervideo [8], and user-selected view [9] were tried (see Section 1.8.2).

The use of digital technology in delivery, whether by cable, satellite, or over-the-air, dramatically increased the quality and number of services available while opening the way for a wide range of innovative new ones, such as high-speed data delivery, interactivity, and narrowcasting of niche services. The use of digital encoding and data compression technology offers up to 200 different channels to TV viewers. In addition to greater channel capacity, digital television offers the opportunity to support additional technological developments, including support for higher resolution, wide-screen formats, and EPGs.

Figure 3.1 shows the approximate dates various technologies were introduced in broadcasting and related sectors. It is clear that the pace of technology had been very slow in the early years of the century, when technologies have lasted a very long time. However, in the 1990s, the pace has been accelerating.

However, various countries around the globe have adopted different systems creating problems of standardization. The main three digital systems available in the world are given in Table 3.1.

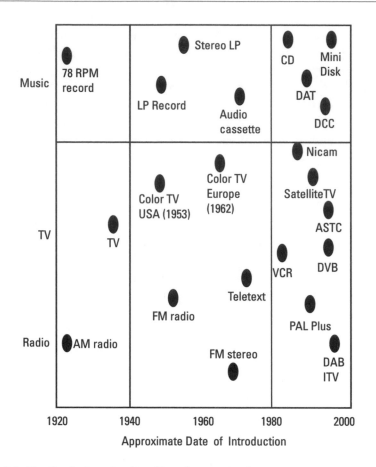

Figure 3.1 Timeline for introduction of broadcast technology.

3.1.1 Digital TV Receivers

Analog television receivers are incapable of receiving a digital signal. Digital receivers have only recently appeared in the consumer markets of some countries. Currently, operators that provide programs transmitted in digital mode also provide consumers with a set-top unit that converts the digital signal so that it might appear on the viewer's analog screen.

The use of digital technology in the receiver offers the flexibility and processing power to decode and display services from a range of sources on wide-screen displays—whose performance is independent of the transport scheme—and to include interactivity and conditional access. The digital TV

Table 3.1
Various Digital Broadcast Systems

Signal	System/Modulation	Region of Adoption
ATSC	8VSB	North America, Taiwan
DVB	COFDM	Europe and 31 countries
ISDB-T	COFDM	Japan

signal remains perfect, as long as the signal is available. This is in contrast to an analog signal, which is subject to a weak signal and static if the signal weakens after a distance (Figure 3.2).

3.1.2 Compact Disc

Digital signals have a much greater resistance than analog signals to noise and interference and can be easily manipulated by using advanced digital electronics. Digital technology offers the viewer far better quality pictures and sound. The CD is a good example. Music on a CD is stored digitally. A CD player reads the binary information from the CD and converts it to sound. A CD player can produce better audio quality than can be achieved by vinyl records or cassette tapes, which use analog technology. Digital sound recording eliminates the hiss and crackle that is a feature of records and tapes.

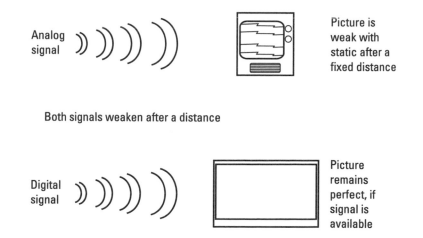

Figure 3.2 Digital TV reception.

Digital television offers CD-quality sound options equal to the cinema's "surround sound."

3.2 ITV

ITV content has developed slowly over 30 years, using primarily analog technology, the telephone, and teletext. The teletext service, where vertical blanking space is made to transmit text in the form of digital data, was introduced in Germany in 1970. An encoder is used to insert digital data and the text is transmitted along with the TV signal. At the receiving end, a decoder helps recover the digital data and display the text on the conventional TV receiver. The service suffers from the low data rate of digital signals [10].

In the 1970s, digital recording technology arrived. One important achievement was the development and deployment of videotext-distributed data networks for general consumption. Videotext is the generic name given to the range of communications technologies based on a similar display format. That is the 40-column, 24-line mosaic graphic display, supporting 8 colors (red, green, blue, cyan, magenta, yellow, black, and white), produced by 7-bit ASCII style coding. Three different strands of communication, using a similar display format and known collectively as Videotext [11], was developed. These were a passive, vertical blanking interval (VBI)–transmitted Teletext [12]; a similar passive product transmitted by cable, called Cabletext; and an interactive system transmitted and received via the telephone network, called Viewdata [13].

This technology was not often linked to television, however. Early systems such as Qube did successfully integrate the technology, but costs were too high for all concerned. In parallel, video conferencing was developed and became a successful business tool. However, a related technology, videophone, was not successful.

ITV reemerged as a strong vision in the 1980s as digital multimedia technology started to be developed. Many telephone companies and cable companies led consortia of technology and service companies to design and market test broadband digital systems. The main aims of these trials were as follows:

- To provide VOD, which requires complex video servers, large bandwidth, and sophisticated home decoders;

- To provide transactional services, such as shopping and banking, that require secure transaction systems, electronic money, and links to the computer systems of collaborating companies.

While many of these trials ostensibly failed as commercial ventures, they have been an important learning opportunity for all involved. Failures as well as successes are important in the development of technology. They make companies and people reassess interactive television, the broader technological issues, and the technology's usefulness.

These trials have resulted in the emergence of intermediate technologies, albeit all of which fall short of fulfilling the vision of full ITV. Levels of video interactivity are shown in Table 3.2. Irrespective of the level of interactivity,

Table 3.2
Levels of Video Interactivity

Level	Description	Applications	Forward Channel	Return Channel
1	Viewer calls up service provider	Pay-per-view subscription TV	Broadcast network	Used only for ordering a service
2	Pseudointeractivity	Teletext, broadcast Internet, local games	Broadcast network	No return channel; information stored locally on the TV
3	Basic remote interactivity	Impulse PPV, VOD, home commerce	Broadcast network	Wireless return: GSM[1], DECT[2], UMTS[3]
	(narrowband)	Internet	—	Integrated return path: cable DTT, UHF antenna
4	Option to use the return channel for receiving video (wideband)	VOD: Internet video streaming, videophone, two-way video	Switched network	Telephone upgraded to: ISDN/ADSL/UMTS
5	Full-service network (broadband)	Full two-way video; telemedicine, teleconferencing	Switched network	Cable modems, VSAT satellite

1. GSM (Global System for Mobile Communications—integrated digital cellular network system) [14].

2. DECT (Digital Enhanced Cordless Telecommunications—a flexible digital radio access standard for cordless communications in residential, corporate, and public environments) [15].

3. UMTS (Universal Mobile Telecommunications System—"IMT-2000" vision of a global family of third-generation mobile communications systems) [16].

however, these are allowing consumers, program producers, and other businesses to catch up in the innovation of content, organization, and media use. In Europe and Asia, broadcasters are already providing enhanced TV information, news, sports, and entertainment. Many others are in the offing. Dozens of studios, networks, cable organizations, equipment makers, and tool providers have teamed up to deliver programming and enhanced content and links to both today's analog systems and tomorrow's digital infrastructure. The enhanced television movement has developed a number of technologies. Pioneer and Toshiba have developed a service called Wink, Intel has developed its Intercast system, Oracle has developed NCI, and Microsoft has bought, and further developed, WebTV.

Technology for ITV is categorized in terms of the standards that sit on top of the physical network, the bandwidth, communications protocols, architecture, and platforms for distributed multimedia. Databases, interfaces, servers, set-top boxes, and secure transactions systems have all developed as key technologies in ITV. The authoring system—the technology that allows content producers to make interactive television—is another important aspect of technology. Many versions of these technologies exist and work in labs. Many have been used successfully in various field trials. The biggest obstacle, however, is to scale them to the mass market, develop common standards, and find combinations that make sense economically to consumers and investors.

A few important underlying ITV technologies have advanced in the marketplace that enable producers to "author once, broadcast everywhere," the professed goal of industry developers. Those are Hypertext Markup Language (HTML)–based 1.0 specifications developed by members of the Advanced Television Enhancement Forum (ATVEF) [17], the Digital Video Broadcasting–Multimedia Home Products protocol (DVB-MHP) [18], the JavaTV application programming interface from Sun Microsystems [19], and XML for controlling metadata [20] supported by the TV Anytime Forum [21]. However, none of these has become a worldwide standard. Companies in Europe and American firms such as Liberate [22], Sun Microsystems [23], OpenTV [24], Excite [25], ICTV [26], PowerTV [27], Microsoft [28], along with interoperability organizations like CableLaboratories (CableLabs) [29] are promoting discussions and organizing workgroups to develop a world standard. Meanwhile, such technologies are finding acceptance in different countries. In Europe, companies have embraced DVB-MHP. ATVEF has found favor in North America. Japan is trying BML. XML is just now gaining real popularity within the Internet community, which may carry over into ITV. Unless a standard can be agreed on,

ITV technologies and, therefore, content programming, will be expensive to produce.

This chapter will look at the various technical systems proposed or in use in interactive television, developments in the underlying technologies, the standards and their producing bodies, such as DVB and the Digital Audio-Visual Council (DAVIC) [30], and the changing fortunes of different technical configurations in the producer and consumer marketplace.

3.3 Building Blocks of ITV

In Chapter 2, we discussed a live experience of an ITV program. In order to provide the services mentioned therein, the following building blocks are required:

3.3.1 Content

Content consists of any form of source material (e.g., movies, games, news, images, and sounds) that will appear on the user's television or PC screen. Vast amounts of material exist already in the form of films, television shows, computer games, and programs. Basic content in any case will be required.

For ITV, the media enhancement is required. The infrastructure's components, such as the Navigation Systems, will involve sophisticated use of all the available media to provide, for example, entertaining but useable menu systems for the user. New authoring tools and a new breed of graphic designers and programmers are necessary to produce these systems. Similar tools and skills will be required to produce program content where the program is not just a simple film or television show but a new concept such as interactive home shopping.

Most level-one television today is based on analog broadcast systems. However, it is inevitable that the future of broadcasting will be digital. Many of the new services can only be achieved effectively by using digital technology. The United States has set 2006 as the deadline for all analog broadcast services to be converted to digital. The date for Europe and Japan is 2010. Many countries are reluctant to announce a firm end date due to uncertainty about the progress of digital television and the value of the spectrum. Nevertheless, many countries are targeting 2010 to terminate analog systems.

3.3.2   Compression

For digital systems, encoders are required to convert analog signals from cameras or other video sources to digital and store them in a highly compressed format. Television signals have substantial redundancy in them, which means that one frame looks much like the next in many cases. This is called temporal redundancy. Even within a frame, the video samples are, to some extent, predictable. This is called spatial redundancy. Eliminating temporal and spatial redundancy in television results in fewer bits to be transmitted. This is called compression of video signal. Consumers and businesses are already using many techniques to compress digital video, for example, the H261 standard (for video conferencing). In the future, fractal and wavelet technology may be used to achieve even greater compression. The key compression standard at the moment is the ISO/MPEG standard.

MPEG was formed in 1988 under the ISO and the IEC to develop standards for digital video and audio compression (Section 1.4.2). MPEG seeks to define standards for the techniques used in achieving digital compression and the way the output binary information is arranged. MPEG has developed formal liaisons with groups within the radio communications and telecommunications arms of the International Telecommunications Union and also works in close cooperation with the Society of Motion Picture and Television Engineers, the FCC, and the European Broadcasting Union.

MPEG-1 [31] is a well-established industry standard. MPEG-1 uses about 1.5 Mbps to achieve slightly better than VHS quality (352 pixels by 240 lines) (see Section 1.4.2.1). The speed is important, since it is the speed which a standard CD-ROM player can achieve and which can be easily transmitted over ordinary copper wire telephone lines using an asymmetric digital subscriber line (ADSL) [32].

The ISO/MPEG-2 [33] standard is now universally adopted. The standard offers data rates of up to 60 Mbps suitable for HDTV. MPEG-2 also defines the protocols for the transport stream, which are important when the data is to be transmitted rather than just stored on a CD. Other features of MPEG-2 include the definition of seven audio channels that could be used for multilingual soundtracks. By using 3–10-Mbps bandwidth, ITU-R-601 [34] quality can be achieved (see Section 1.4.2.2). Extension standards of MPEG include PC-like applications and interactivity. MPEG-4 provides a very low bit rate encoding designed for applications such as low-resolution video telephony [35] (Section 1.4.2.3). MPEG compression requires significant processing power. To achieve full MPEG quality requires powerful specialized processors. On the other hand, decompression is less demanding and there are adapters available for ordinary PCs, which do an excellent job.

MHEG [36] is the standard interactive application programming interface (API). MHEG-5 [37] is a multimedia representation language to interchange units within or across systems (i.e., storage devices to telecommunications networks) (Section 1.4.2.6). MHEG defines specialized links between multimedia files that need no additional processing before they can be used. MHEG-5 is well suited to applications such as VOD and ITV. The advantage of MHEG over HTML [38] for representing multimedia content is the relatively small footprint and memory demands, since a set-top box has limited memory.

3.3.3   Storage Hierarchy and Control System

One of the most important elements of ITV is the means and methods of storing and distributing information and media data. Even after compression, videos require enormous amounts of storage space. Therefore, they are stored on a media or video server. The term "video server" is the central system on which the media is stored. This could be anything from a single PC to a large mainframe capable of serving many thousands of concurrent interactive sessions. But a sophisticated system may well involve massive memories to hold critical components, massive high-speed data banks for data currently being used, and archival storage such as automated tape libraries for older or less popular material. The volumes involved may be many hundreds of terabytes. A Redundant Array of Inexpensive Disk (RAID) is the commonly used storage device for fault-tolerant systems [39]. Broadcast applications demand multiple simultaneous channels accessing the same material and data protection. Disk arrays with RAID satisfy both these requirements.

The server is basically a fast switch with direct access to all this storage. The number and type of processors involved depends on the anticipated load and service level required. Furthermore, these systems may be split across multiple locations, and copies of popular material may be held at many sites.

Managing this video server requires extremely sophisticated software and access techniques. Data needs to be shared and exchanged between the local sites and the broadcast center and between the various storage hierarchies. There is also the challenge of providing an isochronous data stream, which must be provided at a specific rate. If the rate falls too low, then the recipient will not get a smooth video reception. Multiple users accessing different parts of the same video on a single disk offer a significant challenge that is being addressed by techniques such as fragmenting or replicating each video across multiple drives.

In addition to all the software needed to optimize the storage, maintenance, and access to these enormous video databases, there is also the requirement for a control system to service user requests and interface to the storage hierarchy, the transmission system and the subscriber management system.

3.3.3.1 Storage Architecture

A revolution is under way in storage architectures, marked by a continual growth in both physical bandwidth and data speeds for both storage devices and network topologies. The need for faster networks became most evident as the development of faster storage interfaces addressed the demand for more storage. In 1992, the predominant storage interface was SCSI, with speeds at just slightly under 15 Mbps. By 1994, fast-wide SCSI had emerged, moving the disk bandwidth forward to 20 Mbps. Ultra SCSI in 1996 doubled that bandwidth to 40 Mbps, and within two more years, fiber channels emerged, matching storage bandwidth and network data rates at just under 100 Mbps aggregate data rates. The next step for storage interfaces over fiber channels will be 200 Mbps, although in similar time frames, network architectures have already shown that speeds close to 1,000 Mbps are possible over a 10-Gbit Ethernet (10 GbE) (Figure 3.3).

The evolution of storage architectures is pushing the storage model out of the application server domain and onto its own networked storage domain. In earlier times, the model was referred to as server-centric, whereby the file system resided on the application server and data was stored away from the server or file system on just-a-bunch-of-disks (JBOD).

3.3.3.2 Storage-Centric Versus Network Centric

Under the storage-centric model, the usable storage can still grow quite large, but the processes to move, copy, and protect that data depend greatly on the application server or file system. The result is that server performance suffers whenever file system activities become intense. This has led to the evolution of the network-centric architecture. In this model, the file system can decouple itself from the application server. There are multiple application servers, which are connected to the file system and RAID storage system over faster, more efficient topologies. These communicate with the shared storage over a modern high-speed network, which operates at 100 base-T-to-GbE rates. At this point, the network becomes the platform and, in turn, independence between servers and storage types is possible (Figure 3.4).

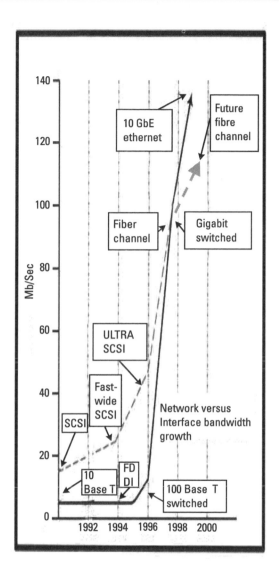

Figure 3.3 Network versus bandwidth growth.

3.3.4 Transmission System

The transition from analog to digital will not inaugurate the "digital super-highway." In practice, digital transmission will not, by itself, change the essential nature of broadcasting. Over-the-air broadcasting will remain,

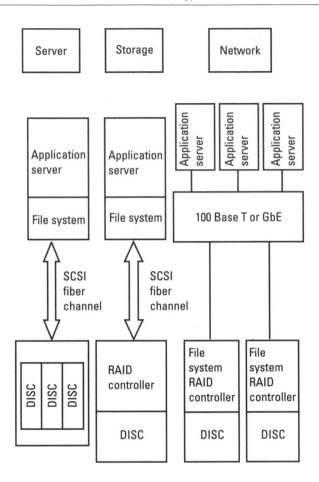

Figure 3.4 Storage architectures.

inherently, a passive, one-to-many, one-way communications system. Figure 3.5 shows alternative means to deliver and receive real-time multimedia content, in addition to the usual broadcast delivery systems (namely terrestrial, satellite, and cable). However, traditional broadcasting will become interactive. Digital technology will multiply the number of available broadcast services, including the following:

- Free-to-air services;
- Subscription services;
- Near-audio-on-demand services.

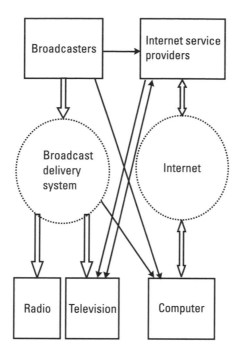

Figure 3.5 Broadcast delivery channels.

Broadcast infrastructure can fully integrate into the Internet's infrastructure and offer "ready to go" bandwidth of up to 150 Kbps that is particularly suited to UDP/IP multicast applications and to "push" applications where a unidirectional stream can efficiently replace a large number of Internet Protocol (IP) unicast streams. The most fundamental protocol is the IP, which is not responsible for transmitting each chunk of data from one system to another. The common term for a network location is "address," and each system on the Internet has an address, which is called an IP address. The Transmission Control Protocol (TCP) provides a virtual connection between two systems, along with certain guarantees on the data chunks (called packets) that are passed between the systems. The User Datagram Protocol (UDP) sends data one chunk at a time (called a datagram) to the other system but does not provide a virtual connection like TCP. UDP also does not provide the same guarantees that TCP does, which means that datagrams may be lost or arrive out of sequence. Each received datagram is checked for internal integrity (like TCP), but if it has been corrupted it is dropped rather than re-transmitted (as TCP does) [40].

Anything distributed via the Internet can be distributed via broadcast television—either via the VBI lines in the case of analog or via MPEG-2 packets in the DTV transport layer.

For video, the transmission must not only be at a sufficiently high rate but must also be delivered isochronously. There are a number of mediums for transmission: twisted pair (namely copper telephone wires), coaxial cable, fiber optic cable, satellite, and microwave. Different systems have different limitations, which we will consider in turn.

3.3.4.1   Satellite

Satellite is being used for many broadcast applications, such as DBS and DTH. The advantage is a large footprint (the area of the earth where a satellite signal can be received) and, thereby, a massive coverage area. However, there are some interesting challenges for fully interactive systems where every user requires a different signal to be sent to his television. Another key problem is latency—the delay inherent in a satellite-based system due to the time it takes the signal to reach the geostationary satellite and then return. Extensive work is being done in this area, with new methods of data packaging being developed in an effort to reduce the delays during an interactive conversation. Satellite transmission is also (as far as domestic use is concerned) a one-way system. The ordinary telephone has to be used as a return path to communicate with the control system. Experiments are being made to provide return channels using a Ka-band uplink to the same satellite with a very small power transmitter for the user.

3.3.4.2   Twisted Pair

This is the most common wired system currently present in millions of telephone lines going to houses. It is also the most limited in bandwidth. However, the upgrading of backbone networks has meant that twisted pair's limitations are often only in the "last mile" as the wire enters the house, and improvements in technology are enabling the necessary speeds to be transmitted over that last mile. The asymmetric digital subscriber loop (ADSL-1) allows 1.5 Mbps (i.e., MPEG-1) to be transmitted into homes and provides a low-speed (64 Kbps) return path. ADSL-2 also allows MPEG-2 transmissions up to 6 Mbps.

ADSL can be used for most common ITV applications, such as Internet access (12 times faster than a 128-Kbps ISDN connection), VOD, and teleconferencing. A very high digital subscriber line (VDSL) is a more advanced version in the pipeline, with speeds of up to 51 to 55 Mbps over 1000 feet.

ASDL's attraction is obvious. It will allow telecommunications companies with twisted-pair networks to provide high-bandwidth ITV services over cable without needing to replace the existing wiring.

3.3.4.3 Coaxial and Fiber-Optic Cable

A coaxial cable can provide 100 channels, each of which is effectively a 36-Mbps pipe. These can be further broken down into 12×3 Mbps MPEG-2 digital television channels, thus giving a total of 1,200 channels (plus spare capacity for control and management). There are many variations on this calculation, but all indicate an enormous number of channels. Likewise, a fiber-optic cable can provide up to 150,000 times the capacity of a twisted pair.

However, current analog cable systems are used mainly to broadcast signals. To become fully interactive requires switching systems that allow true one-to-one transmissions to take place. Telephone systems have the advantage of already being designed with such switching systems.

Another big difference between telephone and coax is the status of existing networks. In the United Kingdom, a single telephone company services virtually every home, while less than 10% of those homes have cable television. In the United States, more than 90% of homes are within easy reach of a cable television system (over 60% are already connected) but the telephone network is owned by a number of competing companies (albeit, regulated by the FCC). In India and many Asian countries, telephone has been the monopoly of the state, while multiple private operators provide cable TV. Each country has its own profile of existing services, complex local regulations, and government interest in the various industries involved, all of which are competing for the lion's share of these future marketplaces. This adds further social, political, regulatory, and economic complexity to the technological challenges. All of these transmission systems are being used in trials today, and it is difficult to predict which will be standard for ITV systems.

In the coming years, broadcast delivery systems will be a mix of conventional broadcast transmitters, Internet, HFC (Section 1.6.3), and satellites. Each delivery system has its own particular strengths and weaknesses. It is clear that there will be no universal solution. Every broadcaster will need to choose the most appropriate delivery mechanism for its target customers. However, as consumers are certain to demand increased portability and mobility, delivery of services from terrestrial radio transmitters will remain an essential element of broadcasting. Another certainty is that we will not converge on a single delivery mechanism. There will be many competing technologies to deliver multimedia services.

3.3.5   Return Path

In a fully interactive system, there needs to be a signal going back from the user to the control system carrying the user's requests. The return path for most ITV applications does not need to be very fast (64 Kbps is quite adequate), since it is transmitting short bursts of control information rather than the much larger media components that are traveling to the home along the main path. Few transmission systems are currently capable of providing a return path. In traditional broadcast-type systems, the return path can be a telephone line or an Internet connection.

3.3.6   Set-Top Box

Future televisions will require more intelligence than current TVs. Until then, an addressable communications box is needed to decode the signals as they arrive at the television. Depending on the system used, it may also need to perform functions such as decompressing the digital signal or handling the return path. The device used for this purpose is known as the set-top box. There are many varieties being used in current trials. New ones being developed now incorporate very significant computing power. Unfortunately, set-top boxes with full MPEG decompression capabilities are currently expensive. However, standardization and mass production will reduce the price.

For most people currently subscribing to cable, the set-top box is very familiar. It sits on top of the TV set, bringing in 50-plus cable channels to the home through a cable wire. The information is mostly one-way, with the TV signals coming in to the home but no consumer response going out. In contrast, a digital box holds a computer chip inside that makes it "smart." This makes two-way communications possible, allowing viewers to do everything from selecting programs on-demand, to banking or shopping directly through the TV, to choosing camera angles or getting additional statistics during sports games. Current estimates suggest that about 6.5 million units were available at the end of 1999, with forecasts that 21.9 million will be on the market by 2003 [41].

One of the problems with digital set-top boxes is that to work to their fullest capacity, they require the use of digital cable. This involves upgrading the existing cable infrastructure to replace the copper wire cables that have been in use for up to 40 years with digital, fiber-optic cables that have far greater bandwidth to bring more information through the pipeline. In the intervening period, only some digital capabilities are possible. Another issue for digital set-top boxes is that no standards currently exist, which means that different manufacturers are producing different versions that may not be compatible with other services or systems.

3.3.6.1 Cable Modems

This technology marries two key appliances: the television set and the computer. Cable modems are being deployed primarily to enhance computer use rather than offer computer functionality in the television set. The cable modem provides high-speed data to the computer through two-way cable wires instead of standard phone lines. The speed of data going through the cable modem is about 100 times faster than in a standard 56K modem. According to the Yankee Group, there are currently about 1.85 million cable modems in use in the United States, with projections close to 10 million in use by 2004.

Because all cable modem users share the same signal path, the speed of data transmission via a cable modem depends on the number of other users. With more users, the demand on the pipeline increases and the speed for each individual slows down. Nonetheless, multiple services operators (MSOs) are offering cable modem service. In the United States, AT&T Broadband Services (which bought out TCI and MediaOne) offers the Excite@Home service, while AOL Time Warner offers a similar system called Road Runner.

3.3.6.2 Smart Card

As analog TV signals fade out over the next 10 years, cable, satellite, and terrestrial TV network operators will use digital bandwidth to deliver more and more interactive content to the sofa-bound set, ranging from purchasing pay-per-view programs and pizza to gambling and impulse buying during commercials. And, in most of these transactions, a smart card will be involved, either as a token that identifies subscribers to the network or as a payment card that completes the purchase. Many of the set-top boxes will come equipped with a second smart card reader, which can accept chip-based debit, credit, electronic purse cards, or loyalty smart cards. The digital set-top boxes already contain a smart card to identify subscribers to the operator. Set-top boxes also use electronic purse cards and have been a focus of such e-purse software vendors as London-based Mondex International and Proton World International in Brussels. In Norway, Bankenes Betalings Sentral AS, BBS, which processes 70% of payment card transactions, launched with its owner banks a roughly 30,000-card pilot set-top box with digital TV operator Canal Digital. Subscribers can pay for programs with the cards.

The FCC has mandated that, after 2005, all digital TV set-top boxes for cable and satellite service contain a chip-based identity token that will allow consumers to buy set-top boxes from electronics stores and other retailers, instead of being tied to the decoder boxes supplied by TV operators.

Motorola Inc., which produced about 85% of the 5.5 million digital cable boxes now in U.S. households, will ship dual-slot set-top boxes that will let consumers download applications onto smart cards over the Internet or service providers' networks. Visa U.S.A. has joined with Motorola to design multiapplication software for the readers.

Open, which is co-owned by BSkyB, HSBC Bank, British Telecom, and Japanese electronics company Matsushita, keeps track of its viewers, or customers, with conditional access smart cards, which are encased in digital decoder boxes the satellite provider distributes to subscribers. SkyDigital stores information about subscribers, including which channel packages they bought using the cards. The cards also authenticate subscribers for Open shopping, banking, e-mail, and other services. In addition, each set-top box comes equipped with a second slot, which Open can program to accept the chip-based debit and credit cards being rolled out across the United Kingdom, or other smart cards, including electronic purse cards. Open will eventually make use of that second slot, which may coincide with the rollout of such new services as allowing customers to load theater or airline tickets or loyalty points onto smart cards.

3.3.6.3 Layers of Software

Set-top boxes have an operating system (OS) that allows the device to boot up and tune channels. Other optional software also exists. GI is incorporating elements of Windows CE in its boxes, while SA, Pioneer, and Toshiba have chosen PowerTV's OS solution. TCI recently decided to use Java as the OS in some of its boxes. The cable TV industry has adopted an Open Cable-compliant set of standards to ensure a common software interface in these boxes. Open Cable offers downloading of software via the network to the set-top box. When the application has outlived its usefulness, it can be flushed out of the system and replaced by another product.

The next layer of set-top box software is called "middleware." Middleware applications on a set-top box are a combination of both content and code. In the era of advanced analog and digital set-top boxes, the software defines the personalization and offerings of the service and the hardware defines certain constraints of the product or service.

Java as a middleware application is playing a minor role in advanced analog boxes, but there is near-universal consensus that its role will grow more important as digital set-top boxes begin rolling out, or at least until a software solution that performs like Java is developed. Java provides open access to the hardware system's capabilities, and that provision will expand the number of application developers that can write to the Java API, which is platform independent and still runs on a variety of boxes.

3.3.6.4   Remote Control and Navigation System

The navigation system allows users to select a service and then specify what particular aspects of that service they require. A remote control (Figure 3.6 and Figure 3.7) is needed to select a service via the navigation system. The request is sent via the return path to the control system, which locates the required data in the storage hierarchy and presents it to the transmission system, which delivers it to the set-top box for display on the television screen. Initial tests by AT&T with 50 families in Chicago in 1993 showed that the navigation system has to be mindlessly simple or it will fail in the mass market.

Analysis shows that to "surf" through 500 channels to search for a desired program takes about 43 minutes. This is the amount of time it takes to manually switch to 500 channels one by one to get the desired channel. By this time, the desired program would probably be over. Surveys show that the average consumer with access to 50 channels only actually uses about 7 of them. An intelligent set-top box can program a user's preferences. However, without facilities like this, along with excellent navigation systems for the interactive services, there will be a plethora of failures.

Figure 3.6   Remote control.

Figure 3.7 KI-W250 wireless keyboard. (*Source:* Sony. Printed with permission.)

3.3.6.5 Advanced Analog or Digital Set-Top Boxes

ITV systems have a very tight integration between the set-top box, the remote control, the navigation system, and the control system. Standards do not yet exist to control this. As the technology matures, standards may be in place that will allow the set-top box to become a standard commodity item that can be integrated into the television itself. To date, MSOs have the option of purchasing two major categories of set-top boxes: digital interactive or advanced analog with digital data capability (Figure 3.8).

GI's most advanced digital set-top box, the DCT 5000, is equivalent to a high-speed network computer (cable modem on board) that delivers EPGs, hundreds of channels, NVOD, VOD (with full VCR functionality), pass-through HDTV programming, and Internet access.

When connected to a two-way 750-MHz cable plant, this device, as well as SA's Explorer 2000 set top enabling many of the same capabilities, takes television into a new direction where it is both used and viewed.

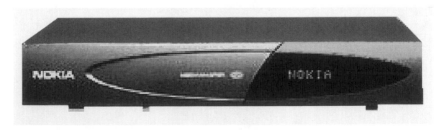

Figure 3.8 Nokia MediaMaster 9800s multiaccess. (Courtesy of Nokia Home Communications.)

Despite all the excitement about digital set tops and what they can do for the television experience, there is still a very strong market for advanced analog boxes, both domestically and internationally.

3.3.6.6 PC/TV or WebTV

The convergence of TV-enabled PCs versus Internet TVs was highlighted with Microsoft's acquisition of WebTV and the inclusion of television capabilities in its Windows 98 operating system. The preliminary Microsoft/Intel specification for PC 99, the minimum system configuration expected for machines, included a TV-tuner card as a built-in option.

Computer vendors are looking at ways to embrace the new digital TV technology. Various companies, including DialTV, WebTV, and Sony, offer set-top boxes exclusively for Internet browsing using the television as the monitor (Figure 3.9). Based on RISC processors, these boxes contain modems to transfer data, interface to PCs and video equipment, and access Web browsers, e-mail, and audio software.

As described in Section 2.7.3.1, there are marked differences in TV and PC reception and they may not merge in the future. PCs may be enhanced to receive broadcasts but may not be liked by many people. Table 3.3 describes the key characteristics of TVs and PCs.

3.3.7 Subscriber Management

Sophisticated systems for administration, billing, and encryption are required to ensure that users pay for the services they use and that copyrights are preserved. Scrambling and conditional access systems ensure that people are paying for what they get. Credit card authentication and payment gateways are used to process purchases made through the television.

Advertisers pay for many television systems. In the future, users will be able to view what they want, when they want, as well as program their set-top

Figure 3.9 Sony WebTV (INT W-250). [*Source:* Sony at http://www.sel.sony.com/SEL/consumer/webtv/products/index.html; printed with permission.]

Table 3.3
Key Characteristics of PCs and TVs

Characteristics	TV	PC
Normal viewing distance	>2m	<1m
Placement	adjacent to external antenna	telephone socket
Sociology	family/household viewing living room/bed room	single user study /bed-room
Psychology	passive/easy to use	active/needs expertise
Services	entertainment, information, education	business, education, games, Internet
Interactivity	remote control	keyboard/mouse
Posture	lean back	lean forward
Screen resolution		
Horizontal	720 samples	1024 pixels
Vertical	576 active lines	768 pixels (typical)
Display	interlaced	progressive
Life span	8–10 years	2–3 years
Price	$300	$800

box to skip all advertisements. Technology will be in place, but it will also open many questions. Who is going to pay for all these new services? Will users need to subscribe or register? Will they pay per use? What do people actually want on their television? How much are they willing to pay for it? Who do they pay? How do all the people who have contributed to the production and other activities get their share of any profits? If the quality is not sufficient (movie watchers will compare it to VHS, games players to 16-bit consoles, and so on), if the price is not right, if the system is not useable, then it will not succeed. This entire area is enormous. The registration, usage, billing, and administrative systems behind future ITV services may well develop into a major industry in their own right.

3.4 System Description for ITV

ITV is, essentially, video broadcasts that incorporate some style of interactivity—be it with data on video, graphics on video, video within video, or retrieving video programming and possibly recording it on a digital hard disk drive for further use. In fact, in ITV, some data is sent along with the program to guide users to switch to the interactivity. The user's response is by way of a return channel. Figure 3.10 shows a normal broadcast system using terrestrial, satellite, and cable transmissions systems. To add interactivity, a return path is needed. This return path may also connect the users to the Internet. Content enhancement, provided by interactive service providers apart from broadcasters, will be required. These providers will play important roles. A generic system reference model is shown in Figure 3.11.

Figure 3.12 and Figure 3.13 show the entire system chain from content provider to customer. Content providers, consisting of video programmers, content creators (program enhancers), and advertisers, provide the enhanced content to service providers. Service providers are responsible for transmission,

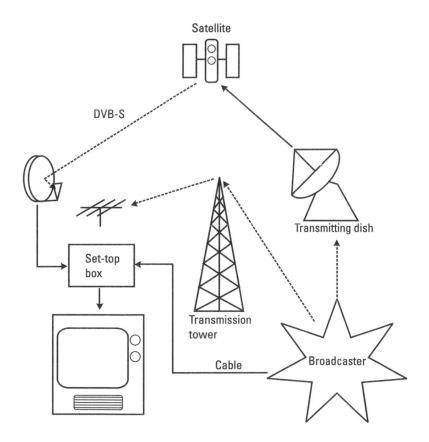

Figure 3.10 Broadcast system using different transmission modes.

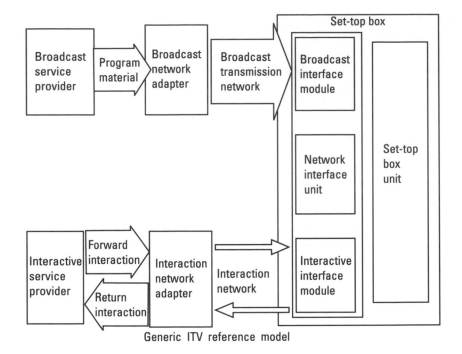

Generic ITV reference model

Figure 3.11 ITV system configuration.

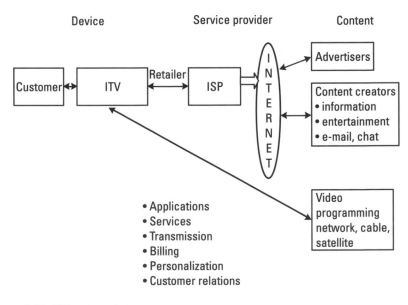

Figure 3.12 ITV system chain.

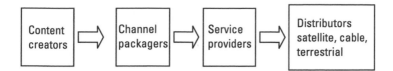

Figure 3.13 Content creation to distribution chain.

billing, personalization, and customer relations. Customers receive the content via the transmission system described above and then interact with the return channel.

Integrated Televisions

First-generation ITV sets are based on television sets currently in production and include a digital terrestrial decoder. These television sets are capable of receiving both analog and digital signals, possibly using a single microprocessor that may combine many functions. These sets are also able to receive enhanced content and on-screen graphics. Internet access on the same TV screen is available. The television sets are based on existing models with added "digital" circuitry. They also contain smart card interfaces for conditional access to subscription products and possible electronic commerce.

Second-generation ITV sets will be capable of only receiving digital signals. They will use a single microprocessor for the whole television. On-screen graphics will use the MHEG-5 standard for Teletext and EPG purposes, and the sets will contain 16:9 interpolators and flicker-free conversion (Figure 3.14).

First- and second-generation ITVs will still need analog compatibility. Both will have modems integrated to embrace Internet TV and home shopping. The next phase will be to provide complete interconnectivity with digital appliances. This will be achieved by the inclusion of the IEEE 1394 high-speed serial bus. Capable of 1000 Mbps, this universal interface will replace SCSI and Enhanced Integrated Disk Electronics (EIDE), allowing set-top boxes to digitally connect to television and DVD-RAMs. Constant increases in processing power and flash RAM will eventually make it possible to transfer some hardware functions to software.

3.5 Internet Protocol Extension for ITV

The architecture of the Web is based on a handful of key protocols, all of which have evolved over time through the open standards process. The

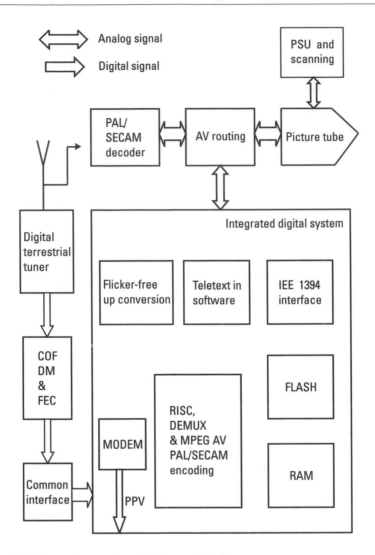

Figure 3.14 Second-generation digital terrestrial ITV.

protocols are sufficiently flexible and extensible to facilitate tremendous technical innovation for their use for ITV.

Several companies have been collaborating to devise a framework to support interactive television content based largely on existing Internet standards. The advantage of this approach is that the resultant platform can support all television environments (analog or digital; cable, satellite, or terrestrial broadcast). This also allows the use of a huge base of tools, media, and know-how that has already been developed for the Web. Existing Web standards, with only minimal extensions for television integration, provide a rich set of capabilities to build enhanced TV content. This approach helps leverage the incredible investment already made in the Web. In addition, this framework can develop as underlying technologies grow, supporting new networks, media, and protocols.

This HTML-based framework has the following building blocks:

- The IP as the primary transport for all data, with multicast IP used over unidirectional links.

- The Session Announcement Protocol (SAP) and Session Description Protocol (SDP) for announcing streams of interactive content.

- HTML and related data formats as the mechanism for describing the presentation of graphics, text, video, and other multimedia.

- Uniform resource identifiers (URI) as the mechanism for referencing all multimedia elements, including television broadcasts and other content delivered over unidirectional links.

- ECMAScript and the Document Object Model (e.g., JavaScript™ and JScript™) for dynamic control of HTML content and the synchronization of HTML content to broadcast video.

- Multipurpose Internet Mail Extensions/Hypertext Transport Protocol (MIME/HTTP)–style headers as the representation for content metadata.

These six elements provide virtually all of the services necessary to design and deliver compelling ITV content. However, these need to be supplemented by the following:

- A protocol to transport content and related metadata over unidirectional links using multicast IP (in other words, an IP multicast-based unidirectional HTTP);

- A specific URL scheme to describe television broadcasts;

- A specific URL scheme to describe content delivered over a unidirectional link;

- A mechanism for sending fragments of ECMAScript over a unidirectional link to specific pages on a client in order to trigger actions that synchronize those pages to the television broadcast.

The above elements and enhancements provide a complete framework for initial deployments and incremental extensions and refinements. These are examined below and are divided into those related to delivery and those related to content design and presentation.

3.5.1 IP-Based Delivery

The IP provides a standard transport mechanism upon which high-level data announcement, delivery, and synchronization protocols can be built. Multicast IP datagrams can be transmitted on unidirectional (broadcast type such as through "over-the air" or satellite) or bidirectional data links (cable). Therefore, higher level protocols built on IP can be used without change on a wide variety of broadcast networks. Standards for transmitting and receiving IP datagrams over virtually every variety of broadcast television network (both analog and digital) either exist today or are fast emerging from the appropriate bodies. For example, the IPVBI working group of the Internet Engineering Task Force (IETF) has published a draft specification to transmit IP datagrams in the VBI of analog television broadcasts worldwide [42].

The delivery of interactive content needs two streams of data. The first stream consists of HTML resources, along with related metadata. The second stream contains small ECMAScript fragments that trigger actions on specific pages of HTML content on the client. With IP protocol in place, both these streams can be delivered successfully.

The first stream, like HTTP (unidirectional), is delivered using MIME-style headers. These headers are used to send attribute-value pairs as metadata. Examples include a `Content-Length:` header to indicate the size of a resource, `Content-Type:` to indicate the media type of the resource, or `Content-Location:` to provide a URL for the resource. The resources, themselves in binary form (and also compressed, as specified by the `Content-Encoding:`), are sent in this stream. Using the "multipart/related" media type, several resources can be grouped together in this protocol and sent as a single unit, allowing large sets of resources to be delivered with "all or nothing" semantics. We will discuss the use of these when we cover enhanced content creation in Chapter 4.

The new protocol has certain enhancements over HTTP properties. The protocol provides some degree of FEC to recover lost or damaged data. This is especially important over broadcast television networks such as satellite, which are prone to very high error rates in certain conditions, and also in cases where there is no return channel. The FEC algorithm can also be defined to allow reconstruction of the data even when the packets arrive out of order. This type of protocol enhancement facilitates efficient reception of the complete content even when the client tunes to the stream in the middle of a broadcast.

ECMAScripts are embedded on HTML pages. These are small programs, often single-method calls. These scripts trigger actions on specific pages of HTML content on the client receiver. Triggers are real-time events delivered for the enhanced TV program. Triggers always include a URL and may optionally also include a human-readable name, an expiration date, and a script. These script fragments are packaged in an appropriate syntax to distinguish them from other multicast packets and, perhaps, describe the page at which they are targeted, or other appropriate metadata.

Announcements are used to tell currently available programming to the receiver. The IP multicast addresses and ports for resource transfer and triggers are announced using Session Description Protocol (SDP) announcements (RFC 2327) [43]. An 8-byte Session Announcement Protocol (SAP) header precedes the SDP header. SAP/SDP notifies clients of the existence of the multicast IP addresses for specific data streams. Both the data and trigger streams described above can be announced in this way. The client (receiver) needs only to listen to an announcement port to discover all available interactive programming. The SAP/SDP packets can contain fields identifying them as announcements for ITV programming, as well as standard SDP fields specifying details of the program, such as language and time span. New fields can be provided to indicate parental guidelines for television content or to distinguish content synchronized to video from other interactive enhancements.

3.5.2 HTML-Based Design

HTML and the related standards that comprise most Web content (such as image formats and style sheets) provide a thorough framework for describing the presentation of multimedia content. The HTML model uses URLs to reference all multimedia data. On the same pattern, two new URL schemes are required: one for television broadcasts and second for other content received over a unidirectional link.

The URL scheme `tv:` is used for television broadcasts. This scheme includes mechanisms for tuning to specific television broadcasts based on station call letters, network name, or channel number. Another URL scheme is used to refer to content received over a unidirectional link that cannot be retrieved on demand. This is essentially a location-independent local name for the data. It refers only to a specific, locally present resource but provides no indication of how to obtain the resource if it is not present. One approach to such a scheme is to create unique namespaces for content based on globally unique identifiers (GUID) and to use shorter names relative to that namespace. Using the relative URL syntax, this allows short, human-readable names to be used to refer to content from within other content, while still preventing name collision between interactive programs. When resources are also available via HTTP (or FTP or similar protocols), a traditional `http:` (or `ftp:`) URL is used to indicate this.

With these two URL schemes, television broadcasts and other unicast content become objects within the HTML model. Other resources (like images or text) can be overlaid on the video using the usual HTML overlay programs. It is possible to create truly ITV-based content by manipulating these objects within HTML. ECMAScript and the Document Object Model provide this capability. Using these standards (as embodied in products like JScript™ and JavaScript™), embedded television broadcasts can be resized on the fly. It is also possible to add, remove, or size the overlays on top of the video.

3.5.3   Transporting Interactive Data

Combined with a batch of new program applications developed by companies like Wink Communications, Intel, Microsoft, Worldgate Communications, WaveTop, and Datacast, the VBI is television's first convergent programming space. ITV involves moving data along with normal program content. Two data channels are typically employed in an ITV connection. One is the VBI and the other is a "back channel," usually a telephone line. The VBI is typically used to send triggers and links. The back channel is used to retrieve interactive content to be displayed on the television screen.

Most broadcast media define a way for data service text to be delivered with the video signal. In some systems, this is called closed captioning or text mode service; in other systems, it is called teletext or subtitling. Some existing broadcast data services provide a mechanism for trigger delivery but not resource delivery, due to limited bandwidth. Content creators may encode broadcast data triggers using these. Broadcast data streams only contain broadcast data triggers, so there is no announcement or broadcast content delivery mechanism. Because there are no announcements, the broadcast data service stream is considered to be implicitly announced as a permanent session.

In 1990, the U.S. Congress passed the Television Decoder Circuitry Act. This legislation requires that all television transmission systems must transmit closed caption data encoded in line 21 of the VBI from its source to the end viewer. While the original intent of the Act was to provide closed captioning for the hearing impaired, it opened up a data path from the broadcaster to the home. Companies like WebTV use this link to send interactive data to their set-top boxes using the Text-2 channel of line 21.

Line 21 of the vertical interval consists of four caption channels (CC-1 through CC-4), four data channels (Text-1 through Text-4), and another service called XDS, or Extended Data Services. All of these channels are multiplexed on line 21. Field one of line 21 is used for CC-1, CC-2, Text-1, and Text-2. Field two of line 21 is used for CC-3, CC-4, Text-3, Text-4, and XDS.

The bandwidth of this channel is limited. The line 21 full-field data rate is 840 bps, or 120 characters per second (7 bits per character). Because captions and text (CC-1 and Text-2) are only in field one, the data rate is half or 420 bps (approximately 60 characters per second). The available bandwidth is divided between all of the services in line 21 on a priority order. CC-1 is highest, and Text-2 is the lowest. The bandwidth used by captions varies greatly depending on the show. Typically, captions use about 25% of available bandwidth, but a very verbose show can take up to 75% or more.

The data formatting, protocol, and channel priority of data encoded in line 21 of the vertical interval are described in the EIA-608 standard [44]. Captions and data encoded in line 21 operate according to Transport A. EIA-608 specifies that closed captioning data takes precedence over all other data. Given the restrictions on available bandwidth in the channel, synchronizing interactive data to the video while giving priority to closed caption data is challenging.

It may help to look at a specific example. Our example will be a game show running on WebTV using Transport A, line 21.

When an interactive program begins, a special trigger is sent using Text-2. This trigger tells the set-top box (STB) that interactive content is available. The STB superimposes the interactive icon on the television screen. When the viewer selects the interactive icon, the STB initiates a connection to the Internet. Once connected, the STB downloads all of the interactive content for that particular show. Once the interactive content is downloaded into the STB, it shifts display modes. The television picture is reduced in size, and a portion of the screen is dedicated to interactive content. From this point on, the viewer uses the controls of the STB to interact with the game and attempt to provide answers along with the show's contestants.

As the show progresses, various triggers and links are transmitted in real time in the vertical interval. These cues change the display and control functions, such as moving on to the next question and answering and revealing answers. Once the STB enters interactive mode, the viewer is essentially looking at a Web page that has been provided over the Internet. HTML, Java scripts, cascading style sheets and other typical Web authoring techniques are used to change the page's content. Other ITV environments may employ different processes.

Transport B

The ATVEF specification has two transport mechanisms: Transport A and Transport B. Transport A was discussed in Section 3.5.3. Transport B is another method of transmitting data in the vertical interval. Transport B can use any available VBI lines. Transport B is for true broadcasts of both the resource data and triggers. As such, Transport B can run on TV broadcast networks without Internet connections, unlike Transport A. An additional Internet connection allowing a return path can be added to provide two-way capabilities like e-commerce or general Web browsing. A combination of standards, Internet requests for comment (RFCs), and trade association specifications set the data rate for encoding data using Transport B, which operates at a higher data rate than Transport A. Transport B has one limitation for the consumer—it will not survive recording on a VCR. Transport A will.

Transport B has a data rate of approximately 10.5 Kbps per line. The ATVEF Transport B has no bandwidth limitations, since the specification is transport independent. However, using Internet RFC 2728, Transport B can be encoded into one or more lines in the VBI using the North American Basic Teletext Specification, EIA-516 (NABTS) encoding waveform.

Transport B is not protected by the Television Decoder Circuitry Act. Therefore, it may not be carried all the way from broadcast stations to homes. In many cities, cable systems maintain ownership of the vertical interval with the exception of line 21. They can strip any data on lines other than line 21 and insert their own data. However, cable systems are under no obligation to deliver vertical interval to the home unless broadcasters have a contract with the system that specifically guarantees this service. Benefits of

Transport B are that it has much higher bandwidth and there is not a requirement to give priority to closed captioning data. As a practical matter, many broadcasters will find that the vertical interval is a very busy place and that multiplexing ITV data on line 21 using Transport A may be the best solution available.

Unfortunately, VBI transport mechanism is a one-way link. It is not possible for the viewer to transmit messages or responses from the STB back to the station using this path. The most common way for viewers to get a return connection (commonly called a "back channel") is to use a dial-up telephone connection to the Internet. This provides a two-way path between the STB and a Web site dedicated to providing interactive content.

3.5.4 Integrating Programming in the VBI

Television transport operators use the standard mechanisms to broadcast data trigger transmission for the appropriate medium (such as EIA, ATSC, and DVB). When the user tunes to a TV channel, the receiver locates and delivers broadcast data triggers associated with the TV broadcast.

3.5.4.1 Data Triggers and Program Enhancements

There are two basic types of program interactivity available using the analog VBI. These are by way of crossover links and real time data enhancements to video programming. A crossover link is a "trigger" sent in a VBI line that takes a viewer from the program to an associated Web site. WebTV Plus subscribers, for example, see an "I" icon on the screen and can click on it. The producers of Baywatch, Disney, N2K (Music Boulevard), and Suzuki (among others) use crossover links. Worldgate Communications uses "hyperlinking," which is the same idea only interactivity is delivered through a set-top box and cable TV head end solution. In the future, all devices (various cable TV set-top boxes, WebTV Plus units and PCs) will read links the same way. The Computer Electronics Manufacturers Association (CEMA) has, in cooperation with Microsoft, NCI (owned by Oracle and Netscape), Intel, and other players, proposed a standard for crossover links called EIA 746 [45], and a working document is currently available.

3.5.4.2 VBI-Popping Screen Overlays

The second level is to send data down the VBI to enhance a video program on the screen in real time. Wink Communications (see Section 3.9.1) and Intel's Intercast (see Section 3.9.2) have proprietary solutions that deliver this kind of enhancement. The advantage with Intercast and Wink is that the data enhancement is on screen with the video. Viewers do not need to go to a Web site, or split the screen (picture-in-picture), to go interactive.

3.6 JAVA Technology for ITV

The digital television revolution is one of the most significant events in the history of broadcast television. Digital television brings significantly improved video and audio quality to the TV experience. It also ushers in the age of true interactive television. As the market for digital television grows, content developers are looking for a feature-rich, cost-effective, and reliable software platform upon which to build the next generation of ITV services, such as EPGs, VOD, personalized TV and enhanced broadcasting.

The Java™ platform, along with the Java TV™ API, provides a development and deployment platform for these interactive services. We discussed Java-based APIs in Section 2.7.1.2. The Java programming language provides content developers with a high degree of control and flexibility over the "look and feel" of their applications. In addition, Java offers security, extensibility, and portability across a diverse array of television receivers, saving content developers time and money getting their interactive applications to market [46]. In Section 1.11, we discussed Java applets for getting users' profiles.

3.6.1 Receiver Using JAVA

The Java TV API is a part of a specification for a broadcast receiver. The digital television receiver gets video, audio, and data from the broadcast stream and processes them through a broadcast media and data pipeline. The receiver gets the media in specific protocols and parses them with an engine specifically designed for each, including the following:

- Broadcast protocol engine;
- Internet protocol engine;
- Video decode engine;
- Software execution engine.

Protocols typically found on a digital television receiver include the following:

- A broadcast protocol stack with flow selection [The OSI Reference Model that defines seven protocol layers is often called a stack, as is the set of TCP/IP protocols that define communication over the Internet. The term stack also refers to the actual software that processes the protocols. For example, programmers sometimes talk about loading a stack, which means to load the software required to use a specific set of protocols. Another common phrase is binding a stack, which refers to linking a set of network protocols to a network interface card (NIC). Every NIC must have at least one stack bound to it.];

- An Internet protocol stack with flow selection (UDP, TCP or a Real-Time Protocol);

- A remote procedure protocol stack;

- A video decode protocol with facilities such as decryption;

- An image decode protocol;

- The Java virtual machine.

The receiver is connected to a broadcast media pipeline, which typically consists of a digital tuner, a demultiplexer, a conditional access module, a collection of media decoders, and a rendering system, through which the media flow. The Java TV API does not require all these components. For example, a receiver may have no conditional access system (if it is not meant for pay channel) or may not have a digital tuner. The Java TV API provides an abstraction that allows the application programmer to remain unaware of the details of the underlying hardware environment. However, a broadcast receiver must always have at least one instance of a broadcast pipeline.

The elements of a typical pipeline and the various stages through which the RF signal is processed in the digital broadcast receiver is shown in Figure 3.15. The key functions performed are given below:

- A "tuner module" is used to tune the RF signal.

- The MPEG-2 transport stream is passed through a demultiplexer and broken into multiple streams (often audio, video, and data).

- The video and audio streams are fed through a conditional access system, which determines access rights and decrypts data.

- The decrypted audio and video streams are fed to a decoder, which converts them into signals appropriate for the video and audio output devices.

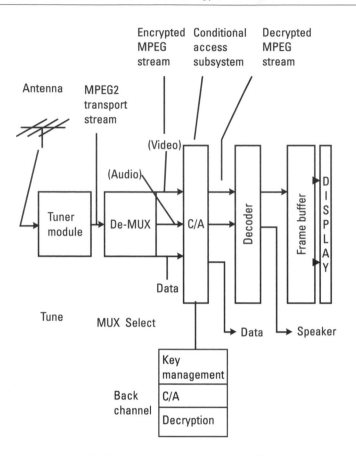

Figure 3.15 Typical digital TV broadcast system. (*After:* Sun Microsystems.)

3.6.2   Software Environment

The software environment on a digital receiver typically consists of the Personal Java application environment, Java TV API, and a real-time operating system (RTOS). The RTOS provides the system-level support needed to implement the Java virtual machine (VM) and Java packages. RTOS and related device-specific libraries also control the receiver through a collection of device drivers.

At the highest layer of the software environment, the application layer, an application can use the Java TV API and the Java packages from the layer below, the Java technology layer (Figure 3.16). The Java applications execute at runtime in the application environment's VM. The Java TV API abstracts

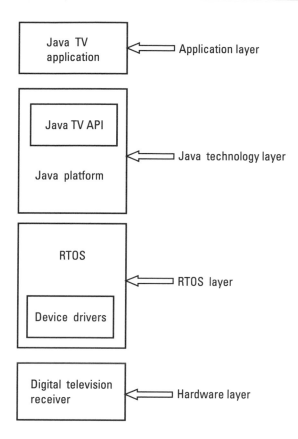

Figure 3.16 Typical software stack on a digital TV receiver. (*After:* Sun Microsystems.)

the functionality exposed by the lower-level libraries to control the television-specific hardware on the receiver [47].

3.7 Competing Standards

Enhanced television is not a new idea. For years, companies have built visions of enhanced television and tried to sell their visions to advertisers and consumer electronics manufacturers. However, none of these proprietary systems caught on due to several reasons. Enhanced television has a "chicken and egg" problem—broadcasters are reluctant to invest in an enhanced television content and infrastructure before consumer electronics companies can provide a reasonable audience size. And consumer electronics companies find it difficult to sell enhanced TV receivers without the support of broadcasters, who must provide compelling and interesting enhanced content.

Today, there are a number of competing standards. ATVEF has developed a set of specifications that will allow content providers to deliver HTML-based enhanced programming to television sets over any means of transport (analog, digital, cable, or satellite) and through standards-based receivers (including analog and digital set-top boxes). By defining the standards used to create enhanced content, the ATVEF specification also defines the minimum functionality required by ATVEF receivers to parse and display this content. One of the ATVEF's major goals was to create a specification that relies on existing and prevalent standards, so as to minimize the creation of new specifications.

The American ATVEF standard is competing with Europe's Digital Video Broadcasting Multimedia Home Platform (DVB-MHP), a TV-centric approach to interactive television (see Section 3.7.1). The Interactive Communicating Application Protocol (ICAP) is a compact protocol developed for today's set-top boxes and TVs that lack the hardware resources necessary to run a full HTML browser. ICAP's compact format also makes it ideal for transmission in the limited data bandwidth of analog broadcasts. Broadcast HTML was created from ATSC-related work to develop the DTV application software environment (DASE). It is a combination of an XML-based subset of HTML 4.0, along with a Java Virtual Machine and Sun's Personal Java API. DASE is a standard being developed for interactive digital TV broadcasts in North America and other parts of the world. DASE will allow interactive and data broadcast applications to be received and executed in a hardware- and software-independent manner on all brands of DTV receivers. The DASE "middleware" consists of Java APIs and XHTML-based on-screen presentation technologies. The solution is harmonized with the Web and takes into account security, conformance, robustness, and TV effects like fade and wipe.

These standards have significant industry support, and none are likely to disappear soon. Many broadcasters are looking to ATVEF and DASE members to provide a minimum level of interoperability between the standards. In any case, customers are looking for their investment in the receiver to survive for 8 to 10 years.

Some companies are not waiting for the standards to settle. CNN, Discovery Channel, and HBO are already delivering enhanced content on a regular or semiregular basis. Consumer electronics companies are designing their next-generation set-top boxes to comply with enhanced television specifications. And embedded Web browser companies are already providing enhancement television support in their browsers. Regardless of what your favorite standard is, you can be sure that enhanced television is coming.

3.7.1 ITV for DVB

The DVB Project is a consortium of more than 250 broadcasters, manufacturers, network operators, and regulatory bodies in more than 30 countries worldwide, committed to designing a global standard for the delivery of digital television. The DVB standard is already well established. Numerous broadcast services using DVB standards with millions of compliant decoders are now operational in Europe, North and South America, Africa, Asia, and Australia.

Equipment compliant to DVB standards dominates the marketplace, and DVB transmissions are on the air in five continents. The DVB system provides a complete solution for digital television and data broadcasting. All DVB systems are based on MPEG-2 video compression and include the necessary elements to bring digital broadcast services to the home through cable, satellite, and terrestrial broadcast systems.

As the DVB Project has progressed, ITV has been identified as one of the key areas ideally suited to an entirely digital transmission system. Many DVB members have developed comprehensive plans to introduce ITV. DVB has worked closely with the DAVIC to develop a set of coherent, practical specifications that meet members' commercial requirements. As for other DVB work, the key has been gathering together commercial requirements to drive the technical work. In the case of DVB-I, these requirements were especially important to help focus the specifications, given the wide diversity of applications that can be envisaged. The result is a set of specifications designed to suit both the needs of the DVB community and the physical characteristics of the individual media. EN 301 192-DVB is the specification for data broadcasting. EN 300 468 is the set of specifications for service information (SI) in DVB systems. The DVB-I system includes the following.

3.7.1.1 DVB Network Independent Protocols

The DVB network independent protocols (NIP) allow session control and protocol tunneling in MPEG-2 transport stream packets.

3.7.1.2 MPEG-2: Digital Storage Media Command and Control

Digital Storage Media Command and Control (DSM-CC) are adopted by DVB to ensure that DVB-I is compatible with internationally agreed standards and flexible enough to meet the challenges of future interactive systems.

3.7.1.3 DVB-IP: Return Channel for PSTN and ISDN

The DVB-IP specification allows interactive DVB systems to operate over existing public-switched telephone networks (PSTN) and integrated services digital networks (ISDN). DVB-IP users are able to select the interactive return channel that best suits their needs and budget.

3.7.1.4 DVB-IC: Return Channel for CATV systems

DVB-IC is an advanced specification allowing full use of community antenna TV (CATV) and cable systems for interactive services. The specification is a practical implementation of the DAVIC 1.x cable modem, allowing return path bit rates of up to 3.088 Mbps.

3.7.1.5 DVB-ID: Return Channel for DECT

The DVB-ID provides a set of return channel specifications using wireless technology that can be used in a variety of systems. The new revised digitally enhanced cordless telephony (DECT) return channel specification contains a vital data service profile for Point-to-Point-Protocol (PPP), which ties in seamlessly with DVB-NIP. With the standard in place, truly interactive services can be provided on mobile phones.

3.7.1.6 DVB-IM: Return Channel for LMDS

The new local multipoint distribution systems (LMDS) return channel specification is based on the DVB-RCC specification.

3.7.1.7 DVB-MHP

The DVB Project unanimously selected Java for its digital ITV broadcasting standard. This agreement led to the DVB-MHP specification, accelerating the availability of rich interactive content running on a wide variety of digital television products from many vendors. The DVB-MHP consists of a software specification for implementation in set-top boxes, integrated digital TV receivers, interactive appliances, and multimedia PCs. The MHP will connect the broadcast television, Internet computing, and telecommunications through these devices and their associated peripherals. Through this agreement, Java technology will be the foundation of the MHP, allowing broadcasters and network service providers to create and offer compelling, cross-platform ITV services.

Java provides an ideal development and deployment platform for interactive services. The Java programming language gives broadcast content developers a high degree of control and flexibility, enabling them to deliver the most dynamic and compelling ITV experiences to their audience. In addition, digital television content written in the Java programming language offers security, extensibility, and portability across a diverse array of television receivers [48].

3.7.2 Digitag

The Digital Television Action Group (Digitag) is a European action group composed of around 50 organizations, including broadcasters, manufacturers of integrated circuits, equipment and consumer products, network and transmission operators, broadcast regulators, and frequency spectrum administrators. Established in 1996, the group aims to accomplish the following:

- Minimize the impact on existing analog services, with which digital services will coexist for many years to come;
- Maximize the number and variety of possible services;
- Optimize coverage;
- Avoid fragmentation of consumer products;
- Provide a quick introduction of interactive services.

One of the group's main activities is to decide on interactive services' development and standardization.

3.7.3 DASE

The emergence of interactive DTV brings a host of exciting opportunities for broadcasters, content providers, tool developers, and equipment manufacturers. Interactive DTV combines aspects of traditional television and the Internet that inspires applications in e-commerce, targeted advertising, VOD, and enhanced viewing services. An enabling technology for applying interactive DTV is being developed by the Advanced Television Systems Committee (ATSC) [49] Digital TV DASE specialist group. Such a standard environment fosters the interoperability concept of write once, run anywhere applications. The DASE API implementation is intended to (1) demonstrate proof of concept of the DASE standard, (2) provide the impetus for conformance testing, (3) aid the design and development of other DASE implementations, and (4) provide an environment to develop and test DASE contents and applications. The reference implementation will serve as the benchmark implementation upon which other implementations can be compared to confirm adherence to the DASE specification. The NIST/DASE reference implementation is a project of the Distributed Systems Technology Group [50] at the National Institute of Standard and Technology (NIST) [51].

The implementation includes the following:

- Java implementation of the API specification;
- Simulated and emulated set-top unit (STU) environment;
- Sample DASE API applications (small test modules);
- User's guide.

A top-down approach has been proposed to implement the API specification. This includes developing Java bindings to the API, developing a simulated DASE environment, evaluating API syntactic and semantic issues, etc. The goal is to develop an API so that the software-only part of the reference implementation can be platform independent. The third phase of the implementation will migrate toward an emulated or real STU environment. NIST is also pursuing collaboration efforts in developing DASE reference applications and conformance tests. A complete NIST "statement of work" document is available [52].

Implementation

Implementation is a tool for demonstrating the feasibility of the DASE API. It can be used to identify inconsistencies, holes, ambiguities, and other potential problems with the specification.

Conformance Testing　　The implementation provides the basis for conformance testing. Thus, the implementation will expedite and facilitate advanced conformance testing procedures.

Prototype Source Code　　All source code and documents are publicly available. The implementation source code can be used as a rapid prototype tool in the design and development of some proprietary DASE API implementations.

Application Development and Testing　　The implementation environment provides a test bed where content and application developers and broadcasters can test their contents or applications on a DASE reference platform.

3.7.4 ATV Forum

The ATV Forum [53] has been formed to build on the work of the ATVEF (Section 3.7.5) and to address the gap between content and technology for enhanced ITV.

ATV Forum Charter The ATV Forum is a nonprofit corporation providing an open industry forum for discussions between content, delivery, and technology industry segments to enable and further the introduction and deployment of enhanced TV content and services initially based on the ATVEF specification.

3.7.5 ATVEF

The ATVEF was formed in 1998 to formulate a solid specification for enhanced programming. ATVEF was founded by 14 industry giants, including Intel and Microsoft, and has since signed on 80 leading companies that support the standard. The ATVEF is mainly focused on delivering interactive services to the varied cable systems in the United States. However, the standard could easily be incorporated on an international basis because it is based on familiar and existing Web standards such as HTML. The ATVEF standard will eliminate the divisiveness inherent in the U.S. cable market that has posed a particular challenge in the delivery of ITV services to broad markets. Few interactive services are able to work "out of the box" with every cable infrastructure or set-top box because there are so many players active in this area. Yet in order to cater to these valuable roaming eyes, cable operators need content that they can deliver to users no matter what mode of transport or receiver technology the viewer has.

The ATVEF Enhanced Content Specification [54] was codeveloped by the ATVEF founders. It is a foundational specification, defining fundamentals necessary to enable creation of HTML-enhanced television content that can be reliably broadcast across any network to any compliant receiver.

The ATVEF industry-wide specification answers the concerns of content providers and device manufacturers that want to assure that they can penetrate the largest share of the market possible. U.S.-based production studios and content providers have been wary of using valuable resources to produce interactive content when they are uncertain of whether or not there will be widely deployed technology to deliver their content to the living room. Similarly, set-top manufacturers have voiced concerns about building set-top boxes that feature powerful processing and expansive memory when there are few interactive applications available to take advantage of these capabilities.

With the ATVEF drawing the industry together, there has been a greater willingness by companies to move forward with their business model to implement ITV services, further relieving any fears that there would be no return on their investments.

As companies begin working together to create interoperable interactive solutions, more services and applications will become available. Collaborations like the one between ATVEF members, Worldgate Communications, and General Instrument will have Internet on TV and EPGs widely deployed in the international market. GI created the SurfView platform, a $99 analog set-top box, that supports Worldgate's Internet on TV services so that viewers can have access to e-mail, channel hyperlinking (connecting to a Web site directly from a television ad or program), and pay per view, among other applications. SurfView is a prime example of technology that helps cable operators satisfy and grow their subscriber base while allowing interactive content providers the chance to reach a diverse market.

OpenTV has been in the middle of this interactive explosion and continues to provide software and professional services to help content providers, cable operators, and device manufacturers deliver interactive services to customers around the world. OpenTV, also a member of the ATVEF, provided enabling software for Worldgate's Internet on TV solution and is currently partnered with General Instrument to help nearly 30 third-party software vendors develop ITV applications at the Acadia Application Integration Center in Boston. Applications that feature VOD, EPGs and Internet services are all being developed with the purpose of delivering on the promise of ITV and reclaiming the attention of viewers worldwide.

3.7.5.1   ATVEF Enhanced Television Standard

The ATVEF is developing standards for creating enhanced ITV content and delivering that content to a range of television, set-top, and PC-based receivers. ATVEF defines the standards used to create enhanced content that can be delivered over a variety of mediums—including analog (NTSC) and digital (ATSC) television broadcasts—and a variety of networks, including terrestrial broadcast, cable, and satellite. A broad framework and elements of innovation for enhanced TV by ATVEF are given in Table 3.4.

Elements of Innovation

- A protocol for transporting content and related metadata over unidirectional links using multicast IP (in other words, an IP multicast-based unidirectional HTTP);

Table 3.4

Broad Framework for Enhanced TV by ATVEF

The IP as the primary transport for all data, with multicast IP used over unidirectional links.
The SAP/SDP for announcing streams of interactive content.
The HTML and related data formats as the mechanism for describing the presentation of graphics, text, video, and other multimedia.
URIs as the mechanism for referencing all multimedia elements, including television broadcasts and other content delivered over unidirectional links.
ECMAScript and the Document Object Model (i.e., JavaScript™ and JScript™) for dynamic control of HTML content and synchronization of HTML content to broadcast video.
MIME/HTTP-style headers as the representation for content metadata.

- A specific URI scheme for describing television broadcasts;

- A specific URI scheme for describing content delivered over a unidirectional link (and therefore not available on demand);

- A mechanism for sending fragments of ECMAScript over a unidirectional link to specific pages on a client in order to trigger actions that synchronize those pages to the television broadcast.

Apart from defining the standards used to create enhanced content, the ATVEF specification also defines the minimum functionality required by ATVEF receivers to parse and display this content. Thus, the ATVEF specification addresses both the transmission and reception of enhanced content. The ATVEF specification relies on existing and prevalent standards so as to minimize the creation of new specifications. Keeping this objective in mind, the group based its content specification on Internet technologies such as HTML and JavaScript. This strategy has multiple benefits. The content creators are comfortable creating content based on existing Internet technologies, the content specifications are fully Web compatible, and millions of pages of potential content already exist. With the availability of many of today's Web-authoring tools, practically anyone can become an ATVEF content developer.

The ATVEF 1.0 Content Specification mandates that receivers support HTML 4.0, JavaScript 1.1 and cascading style sheets. This is a minimum content specification because all receivers must support these standards. However, receivers are allowed to support other standards as well, namely

Java and virtual reality modeling language (VRML). Establishing a minimum content specification ensures that the enhanced content is available to the maximum number of viewers.

Availability of a Web browser on a television set introduces exciting possibilities. The user can be provided exciting new content apart from surfing the Web. To support this, the ATVEF specification has added new extensions to existing standards. The most prominent extension to HTML defined by the ATVEF specification is the addition of a `tv:` attribute that allows the insertion of the television broadcast signal into the content. The attribute is used in an HTML document anywhere that a standard image may be placed. The TV channel can be displayed in a small picture-in-picture screen. Creating such an enhanced content page is as easy as inserting an image in an HTML document [55].

The AVTEF specification also defines the way content gets from the broadcaster to the receiver. The receiver is informed that it has enhancements available for the user to access by sending special information known as triggers.

Triggers

Triggers contain information about enhancements that are available to the user and used to alert receivers to incoming content enhancements. Triggers are sent over the broadcast medium. Among other information, every trigger contains a standard universal resource locator (URL) that defines the location of the enhanced content [56]. ATVEF content may be located locally—delivered over the broadcast network and cached to a disk—or it may reside on the Internet, another public network, or a private network. Triggers may optionally contain a human-readable description of the content. For example, a trigger may contain a description like, "For details of future program, press browse." The receiver will then display the Web page containing future programs. Triggers may also contain expiration information that instructs the receiver about the duration of enhancement availability to the viewer. A checksum attribute within a trigger ensures the integrity of the delivered information. Triggers may also contain JavaScript fragments that trigger execution of JavaScript within the associated HTML page. JavaScripts are used to synchronize the enhanced content with the video signal and update the dynamic screen data.

Transports

Transports are the means of delivering content using only a forward channel or in association with a back channel. The ATVEF specification also defines how content is delivered. A television or set-top box may or may not have a connection out to the Internet. Hence, the ATVEF specification describes two distinct models for delivering content—Transport Type A and Transport Type B. Transport Type A is defined for ATVEF receivers that maintain a back channel or return path to the Internet. Generally, this network connection is provided by a dial-up modem but may be any type of bidirectional-access channel such as cable or HFC. Transport Type A delivers only triggers without additional content. Accordingly, in Transport Type A, all data is obtained over the back channel, using the URLs passed with the triggers as a pointer to the content.

Transport Type B delivers both ATVEF triggers and its associated content via the broadcast network. In this system, the broadcaster pushes content to a receiver, which in turn stores it in case the user chooses to view it. Transport B uses announcements sent over the network to associate triggers with content streams. An announcement is used to describe a content stream. It includes information such as bandwidth, storage requirements, and language. It is important to note that enhancements may be delivered in multiple languages. The receiver, in most cases, needs to store any content that will be displayed. It uses announcement information to make content storage decisions. If a stream requires more storage space than a particular receiver has free, the receiver may decide to discard some older streams, or it may choose not to store the announced stream. A drawback of Transport Type B is that if a user chooses to start watching a show near the end, there may not be time for the content to be streamed to the receiver. As a result, the user may not be able to view some or all of the content. However, the advantage is the push delivery of the streams with the triggers and announcements on the forward channel, without the need of a back channel.

Delivery Protocols

The ATVEF specification also defines a reference protocol stack used for content delivery. All of the high-level protocol layers are well defined for every ATVEF implementation except the protocols for the link layer and physical layer that depend on the broadcast network. This is because it is not possible to transmit analog data over cable in the same way as digital data over satellite. Figure 3.17 shows a standard ATVEF protocol stack for delivery of enhanced content.

For traditional bidirectional Internet communication, the HTTP defines how data is transferred at the application level. However, a two-way connection over a broadcast medium is not possible and, as such, a unidirectional application-level protocol for data delivery is required. ATVEF defines

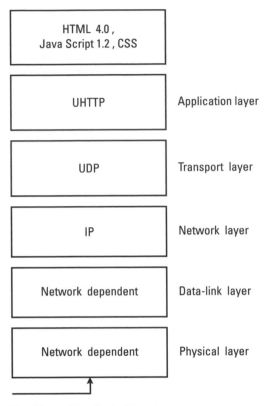

Satellite, cable, off-air video data

Figure 3.17 Standard ATVEF protocol stack for enhanced content.

this protocol as the Unidirectional Hypertext Transfer Protocol (UHTTP). UHTTP is based on UDP, as opposed to TCP. UDP is a connectionless protocol most suitable for a broadcast network. Like HTTP, UHTTP uses traditional URL naming schemes to reference content. Therefore, content creators can reference enhancement pages using the standard `http:` and `ftp:` naming schemes. To this, ATVEF adds the `lid:`, or local identifier, URL naming scheme. The `lid:` naming scheme allows a content creator to reference content that exists locally (on the receiver's hard drive, for example) as opposed to on the Web.

With HTTP, the TCP layer provides well-defined error detection and retransmission facilities. But, for a unidirectional protocol like UDP, there is no possibility for retransmission requests. Thus, UHTTP must implement error correction without retransmission. This type of error correction is called FEC and is used in digital broadcast systems. Using FEC algorithms, the data is regenerated with only the received information. Of course, if the data is too badly corrupted, FEC may not be able to regenerate it. With an emphasis on error correction instead of detection, coding schemes are more similar to the algorithms used in data storage like digital tapes and CD-ROMs than those used in traditional bidirectional communications.

Bindings

Binding is defined as the way in which ATVEF data is delivered over a particular network—from the network layer protocol down to the physical layer. For ATVEF to provide interoperability between broadcast networks and receivers, it is important that every network has only one binding. Each binding must provide a fully comprehensive definition of the interface between the broadcast network specification and the ATVEF specification. ATVEF has defined bindings for delivering data over IP multicast, as well as over NTSC. The transmission of IP is defined (or can be) for virtually every type of television broadcast network. As such, the binding to IP is considered the reference binding.

3.7.5.2 ATVEF for NTSC?

NTSC is the standard for analog television broadcasts in the United States. The NTSC standard defines a frame as consisting of 525 horizontal lines, each line scanned left to right. It has 21 extra "lines" that make the VBI. Only the first nine lines of the VBI are actually required to reposition the cathode ray. This leaves 12 more lines that can be used to broadcast data. In fact, in the United States, closed captioning data has been broadcast on line 21 for many years. Each line of the VBI has a transmit rate of about 17 Kbps. So, in theory, the VBI associated with each NTSC-encoded television channel could carry up to 204 Kbps (12 lines at 17 Kbps/line) of piggyback data. However, after taking into account the overhead associated with the various protocol layers and the need to prevent conflicts with closed captioning and other data already broadcast within the VBI space, the maximum achievable rate for ATVEF data transmission is somewhere around 100 Kbps.

Transport Type A

ATVEF triggers are broadcast in line 21 of the VBI. For purposes of data integrity, the NTSC binding for Transport Type A requires that each trigger contain a checksum. The binding also recommends that the trigger length does not exceed 25% of the total bandwidth of the line, to avoid conflicts between triggers, closed captioning data, and data from any future services that might also use line 21.

Use of line 21 for ATVEF triggers provides advantages for receiver manufacturers. Since most standard NTSC video decoders already have the ability to extract line 21 of the VBI for closed captioning data, hardware manufacturers are not forced to upgrade to more expensive decoders that support data extraction in other lines of the VBI.

Transport Type B

The Transport Type B NTSC binding is used to send triggers on line 21 of the VBI. The binding also includes a mechanism for delivering IP datagrams over other VBI lines. IP over VBI (IP/VBI) is an Internet draft of the Internet Engineering Task Force (IETF). Figure 3.18 illustrates the protocol stack defined by ATVEF down to the IP layer and defined by IP/VBI below that.

The NTSC television standard is at the bottom of the stack. At the lowest level, the NTSC television signal transports North American Basic Teletext Standard (NABTS) packets. NABTS is a method of modulating data onto the VBI. A typical NABTS packet gets encoded onto a single horizontal scan line (VBI line). NABTS, by way of its own forward error correction, can provide for correction of single bit, double bit, and single byte errors, as well as having the ability to regenerate an entire missing packet. The NABTS packets are removed from the VBI to form a sequential data stream. This data stream—encapsulated in a Serial Line Internet Protocol (SLIP)–like protocol—is unframed to produce IP packets, which are handled equivalently across all ATVEF-network types that implement the IP reference binding. ATVEF bindings to every other major video network standard, including PAL, SECAM, and ATSC (digital terrestrial broadcast), as well as cable and satellite, are under development.

3.7.5.3 Design Issues

A specific network binding is detailed enough that anyone implementing a broadcast network or building an ATVEF receiver has enough information to make the design ATVEF-compliant. The first major decision when designing an ATVEF receiver is whether to support Transport A or Transport B. For a satellite television set-top box, Transport B should be chosen since satellite does not provide a return path to the service provider. For a cable television set-top box that doubles as a cable modem with dedicated Internet access, Transport A is the obvious choice. Choosing to support a high-bandwidth

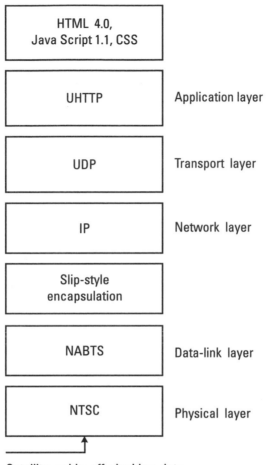

Satellite, cable, off-air video data

Figure 3.18 ATVEF protocol stacks down to the IP layer and defined by IP/VBI.

option like Transport B requires additional hardware and software performance. For a standard NTSC binding for Transport B-NABTS encoding of the
data in the VBI, an NTSC video decoder that will parse all VBI lines in hardware is required. The other option is to do the NABTS decoding in the software. Unfortunately, software decoding is very processor intensive. The user
interface (for example, an indication of enhancements) is another issue in the
design of receivers. A whole set of different issues will need to be addressed for
analog/digital PAL/SECAM receivers.

3.7.5.4   ATVEF Integration with DVB Using IP/Multiprotocol Encapsulation

The purpose of having a definition for integrating ATVEF with DVB is interoperability. The ATVEF Enhanced Content Specification [57] describes how to send ATVEF using IP/UDP (including multicast IP). This specification describes integrating ATVEF with the DVB transport. DVB has defined the mechanism for carriage of IP/UDP via Multiprotocol Encapsulation (MPE) (EN 301 192). The specification does not discuss the upcoming DVB MHP specification.

Transport Binding for ATVEF Using DVB-Defined Transport.   ATVEF Transport B, which defines delivery of announcements, triggers, and content within the broadcast delivery, is the transport variant for the discussion of ATVEF integration with DVB.

IP Transport.   The use of IP (multicast or unicast UDP) is one proven solution to carry ATVEF content. Most modern networks provide support to transmit IP packets. The IP binding of ATVEF can therefore be used to implement ATVEF on DVB.

DVB Transport Binding.   A complete ATVEF service on DVB typically consists of audio, video (DVB-specified), and IP data, which form the service components of a DVB service (Figure 3.19).

ATVEF transport includes the content, triggers, and announcements. When using the DVB binding, ATVEF transport is delivered through IP (multicast or unicast UDP). The software layer diagram is shown in Figure 3.20. IP is delivered in DVB by using MPE of IP as specified in EN 301 192.

EIT, SDT, and ATVEF `Data_Broadcast_Descriptor`.   The Service Description Table (SDT) or Event Information Table (EIT) contain information (such as descriptors) referencing audio-video and ATVEF data streams. Audio-video is associated with ATVEF data via the SDT or EIT as specified in EN 300 468. A string indicating ATVEF content is added to the `data_broadcast_descriptor` for streams containing ATVEF data. It is recommended that the `data_broadcast_descriptor` be placed in the SDT if the enhancement is continuously available for the service. It should be placed in the EIT for ATVEF enhancements accompanying time-limited events, namely a specific TV show. If there is one `data_broadcast_descriptor` referencing an ATVEF stream in both the EIT and SDT, then the descriptor in the EIT takes preference over the one in the SDT.

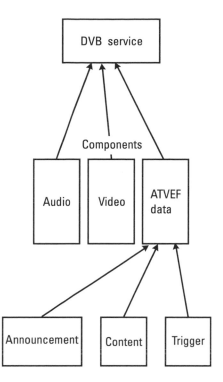

Figure 3.19 ATVEF circuit on DVB.

Number of ATVEF Streams in Service. ATVEF data stream is described with one `data_broadcast_descriptor` per ATVEF data stream. If more than one stream in a service is found labeled as carrying ATVEF data, then data is read from all of those ATVEF source streams. This allows for downstream insertion of ATVEF presentations into a service by using additional streams. While there is no technical restriction on the number of data streams from the sending end, some receiving devices are limited in the number of streams that they can read at the same time. Therefore, streams to send ATVEF data is generally limited to four per ATVEF service. Some systems use a different local interface IP address for each stream carrying IP data. This helps keep datagrams of different streams separate.

Data_broadcast_descriptor for DVB. The `data_broadcast_descriptor` for the ATVEF stream is defined according to EN 300 468 describing an MPE stream (EN 301 192). The `data_broadcast_id` in the descriptor

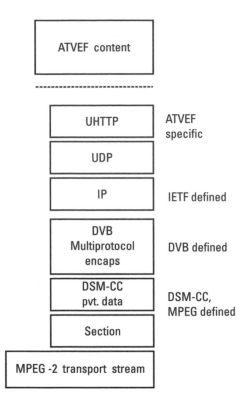

Figure 3.20 Software layer diagram for ATVEF with DVB.

is set to 0x0005 (MPE). Other specific values for some fields of the descriptor are described below.

Other Applications in the Same IP Data Stream.   The ISO/IEC 13818-1 program element conveying the ATVEF IP datagrams may also include other IP datagrams belonging to services other than the ATVEF service. All datagrams are encapsulated in datagram_section as specified in EN 301 192 for DVB data broadcasting. In such cases, the ISO/IEC 13818-1 program element conveying these IP datagrams have an ISO/IEC 13818-1 stream_type value equal to 0x0D.

DVB Details   Table 3.5 describes the mode of filling some fields such as data_broadcast_descriptor, multiprotocol_encapsulation_info structure and datagram section. The term "handled transparently" means that

Table 3.5
Some Fields and Their Values for AVTEF IP on DVB MPE Transport Binding

Data_broadcast_descriptor	Values for ATVEF IP on DVB MPE Transport Binding
data_broadcast_id	0x0005
ISO_639_language_code	Transparent. Uses a language code with ASCII characters.
text_length	The total length of the string.
text_char	Semicolon delimited fields. To indicate thatATVEF data is included, one of the fields must contain "ATVEFx.x" where x.x is the version of ATVEF used. Another one of the fields must contain the string "IP" to indicate that the stream contains IP datagrams. Other applications can have their own strings that appear in other fields. An example string is "IP;ATVEF1.0."
Multiprotocol_encapsula-tion_info **Structure**	**Values for ATVEF IP on DVB MPE Transport Binding**
max_sections_per_datagram	1 (always for IP)
MAC_IP_mapping_flag	1 if data is IP multicast; 0 otherwise. (1 means to do IP to Mac address mapping for IP multicast as in RFC 1112)
Datagram Section	**Values for ATVEF IP on DVB MPE Transport Binding**
MAC address_1to MAC_address_6	Handled as defined by RFC 1112. The IP address is used for IP to Mac address mapping as described in RFC 1112. The different IP addresses of announcement content or trigger stream are reflected in different values in these fields. Since IP addresses are reflected in these fields, receivers can apply filtering.
LLC_SNAP_flag, LLC_SNAP bytes	Flag set to 0 for IP data.
Scrambling bits: payload_scrambling_control, address_scrambling_control	Handled transparently.

any value specified in EN 301 192 can be chosen. The unlisted fields are filled according to EN 301 192 and other applicable DVB specifications.

3.7.6   B XML/BML Draft Specification

The XML Working Group of the Association of Radio Industries and Businesses (ARIB) is finalizing an XML-based multimedia content format that can be commonly used for data broadcasting services of broadcast satellite and terrestrial broadcasting, based on the following objectives:

- It satisfies the requirements for digital broadcasting and its data broadcasting, which have been discussed in the Digital Broadcasting System Committee of the Telecommunications Technology Council.

- It ensures high extensibility, which is a feature of the XML-based coding scheme, and international exchangeability.

- It can be introduced with minimum influence on receiver costs from the start of broadcast satellite digital broadcasting.

3.7.6.1   B XML and BML

An XML application language defined in this specification consists of a set of tags for multimedia content format. The application language is called Broadcast Markup Language (BML) [58]. The scope is the same as that which has been drawn up using previously existing ARIB requirements for multimedia services and those currently drawn up as a presentation engine by ATSC/DASE.

The XML tags defined in each application are provided by the document type definition (DTD) of its application, respectively, and transformed to the BML tags by the Extensible Style Language Transformation (XSLT) for the presentation process of the receiver terminal. XSLT is the language used in XSL style sheets to transform XML documents into other XML documents. XSLT was written and developed by the XSL Working Group and became ratified by the W3C on Nov. 16, 1999. DTD is a type of file associated with Standard Generalized Markup Language (SGML) and XML documents that defines how the markup tags should be interpreted by the application presenting the document. The HTML specification that defines how Web pages should be displayed by Web browsers is one example of a DTD. XML promises to expand the formatting capabilities of Web documents by supporting additional DTDs.

The XML scheme defined with this method is called B XML, for Broadcast XML. B XML is an extended scheme of BML. Figure 3.21 illustrates the

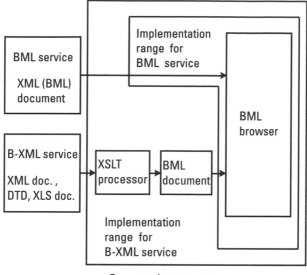

Presentation process

Figure 3.21 BML- and XML-like application using B XML scheme.

relationship between BML and B XML. At the presentation, the B XML documents serve as BML documents.

XSL is a style-sheet language that specifies the presentation method of the XML document. This specification assumes that any XML document is transformed into a BML document with tag-sets of multimedia content format by XSLT.

3.7.6.2 Scope of XML-Based Coding Scheme

Figure 3.22 depicts protocol stacks of data broadcasting. Multimedia services, which are described using the XML-based coding scheme, consist of monomedia representation data, such as bitmap image, audio, characters, graphics, and multimedia representation data, that integrates and controls the monomedia data. Multimedia documents are described with the format using BML or B XML. In the case of broadcasting, content is transmitted according to the DSM-CC carousel or the additionally specified transmission scheme by the future extension. The content may also be provided from the interaction path when there are additional services for broadcasting. The interaction path can be provided in the form of telephone lines, cable, or the

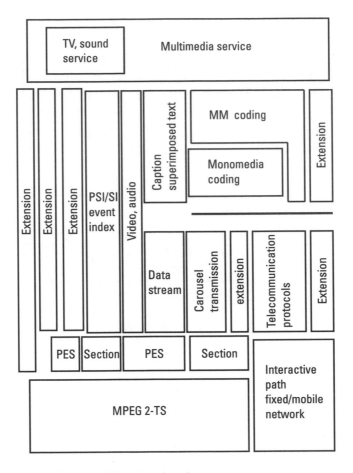

Figure 3.22 Protocol stacks of data broadcasting.

Internet. Telecommunications control via the interaction path is used to send orders and questionnaires from the receiver to the computer center, in addition to the acquisition of the content.

3.8 Interactive Terrestrial TV Integration

The Interactive Terrestrial TV Integration (ITTI) is evaluating a design for a UHF terrestrial broadcast ITV system with enhanced compatibility with other systems (interactive CATV). The system uses a forward channel embedded within the DVB-T transport stream and a synchronous frequency division multiple access (SFDMA) UHF return channel [59]. The work is an extension of earlier ACTS projects, notably INTERACT, VALIDATE, and DVBIRD. The main advantage is that a telephone line is not required for the return channel.

Technical Approach

The ITTI covers three aspects of digital interactive terrestrial TV; namely, services and network architecture, applications, and digital interactive terrestrial set-top boxes (DIT-STB). The DVB and DAVIC requirements for ITV services, and the potential of the DIT-STB, are being analyzed. In parallel, the interactive network architecture is proposed to be defined according to the DVB and DAVIC general model and refined for both the transmitter and subscriber or user site. Finally, radio frequency engineering studies will be conducted based on input from VALIDATE. The project will build and evaluate Java-based applications, with various levels of set-top box intelligence, to demonstrate system capabilities and limitations. It will start with the DIT-STB specification and architecture produced by DVBIRD and INTERACT and consider integrating and interfacing the DVB-T receiver system with the SFDMA UHF interactive system. A system diagram is given in Figure 3.23. It will be observed that the return channel is also in the UHF band.

3.9 A Few System Descriptions

We will now examine two working systems of ITV.

3.9.1 Wink TV

The term best used to describe the service from Wink [60] is enhanced broadcasting. Wink is using the VBI to provide interactive graphics or a data and text overlay on either program content or advertising. The software to enable this service resides at the cable head end, where the cable signal itself comes from. If the programming material is being enhanced, that takes place at the programming network. Wink is not VBI-only technology. It can be delivered in a number of ways (including MPEG-2 video streams), but using the VBI is an efficient way to deliver it to the current generation of set-top boxes. Wink requires one line of the VBI, or a pass-through rate of about 14.4 Kbps. It is designed for a low overhead application and includes FEC to account for signal distortion.

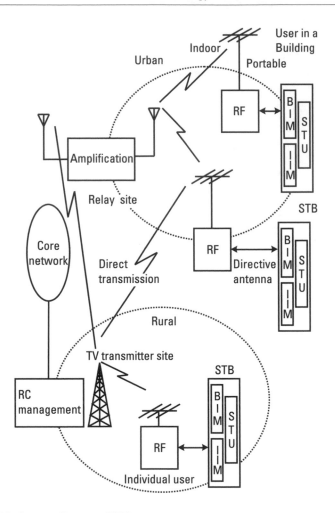

Figure 3.23 System diagram of ITTI.

Wink allows both programs and ads to be enhanced. When the viewer sees a special icon appear in the top left-hand corner, she can click on it to bring up the information overlay on the screen. It replaces the existing screen content, so that the viewer cannot see the ongoing program while involved with the interactive element. However, in an effort to appease advertisers, who are concerned that someone interacting with the first ad in a commercial break would then be unavailable to see the second or subsequent spots, Wink maintains that the interaction itself cannot last longer than 30 seconds. Viewers are not permitted to begin interacting with an ad unless there is sufficient time to complete the interaction. If that is not the case, they are directed to a special, dedicated cable channel where all the Wink-enabled ads reside, so that they can explore them at their leisure.

The system description of an enhanced TV broadcast offering t-commerce is shown in Figure 3.24. Wink is offering this service to about 200,000 consumers in the United States. The various components of the system are described below:

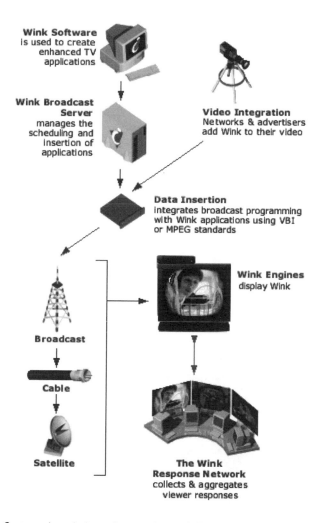

Wink Software is used to create enhanced TV applications

Wink Broadcast Server manages the scheduling and insertion of applications

Video Integration Networks & advertisers add Wink to their video

Data Insertion integrates broadcast programming with Wink applications using VBI or MPEG standards

Wink Engines display Wink

Broadcast

Cable

Satellite

The Wink Response Network collects & aggregates viewer responses

Figure 3.24 System description of an enhanced TV broadcast offering t-commerce. (Courtesy of Wink; http://www.wink.com/contents/tech_diagram.shtml.)

3.9.1.1 Enhanced Programming

A high-level authoring tool is used to create interactive programming and advertising applications. The program is Windows based and graphically oriented (Figure 3.25). It enables the creation of simple interactive applications. A designer simply drags objects from an object palette onto forms to create interactive enhancements. More complex applications can be created by fully using the Basic scripting language to control the behavior of objects and forms. The software allows producers, advertisers, and third-party developers

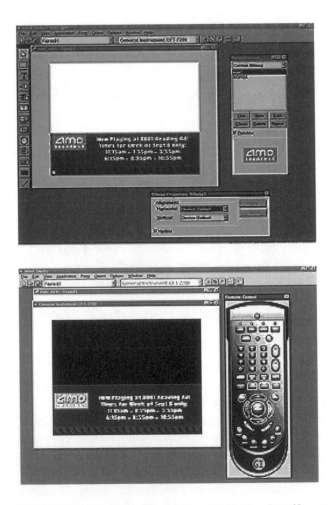

Figure 3.25 Windows-based software for enhanced programming. (Courtesy of Wink; http://www.wink.com/contents/tech_studio.shtml.)

to add elegant interactive overlays to shows and commercials. Live data updates from Web sites and databases are also allowed.

Producers or content providers using Wink have to purchase the authoring tools and server. The MSO must also purchase a Wink data server and download the application to the set-top box, then tune it to the head end. Intercast channels use the VBI to send HMTL files (Web browsing files) to a PC. VBI data is inserted into the signal through a commercial insertion device (such as those from Norpak and others). Intercast channels currently include CBNC, CNN, M2, QVC, Lifetime, The Weather Channel, and HGTV.

3.9.1.2 Broadcast Server

The Broadcast server manages the scheduling and insertion of enhanced applications into television programming. It is designed to integrate with station management equipment such as commercial insertion systems to enable broadcast and cable networks to automate the delivery or interactive enhancements to programs and advertisements. Local network affiliates and cable operators can also add interactivity on a local level using the broadcast server.

3.9.1.3 Client Software

Client software enables the display and generation of viewer responses. Client software can be installed in most new advanced analog and digital set-top boxes at the factory. It can also be downloaded through cable systems, satellite systems, or telephone modems into certain set-top boxes already installed in a viewer's household.

3.9.1.4 Response Server and Response Network

The response server collects viewer response data, which is generated by program enhancements. Responses are retrieved directly from television and satellite set-top boxes through phone dial-up and from cable set-top boxes through cable head-end systems. The response server aggregates the response data and delivers it to the response network, which is designed to enable the collection and aggregation of viewer responses, requests for information, and purchase orders. It then transmits these responses to advertisers, merchants, broadcast and cable networks, and cable and direct broadcast satellite system operators.

When a consumer responds to an enhancement and requests information or orders a product, the client software in the set-top box or television generates a response packet. These response packets are collected, aggregated, converted into full electronic orders (such as name, address, and credit card information, if appropriate), and sent on to the advertiser or network for fulfillment. The response network combines custom data processing solutions with off-the-shelf electronic commerce systems to provide a complete end-to-end solution for customers.

Obviously, the challenge of smoothly handling millions of simultaneous retail transactions in real-time is daunting. A distributed database network is used to cope with the tidal waves of responses that television offers generate. It is all linked directly to merchants and vendors that pick, pack, and ship consumer orders.

Over time, televisions and set-tops are expected to become more sophisticated, and high bandwidth digital transmission will replace analog broadcasts.

Wink-enabled interactive services are projected to reach up to 4 million DISH network subscriber homes by mid-2003.

3.9.2   Intel's Intercast System

Intel's Intercast [61] is really a hybrid between datacasting and Wink pass-through interactivity. The data files show up on the hard drive, and the Intercast client software manages a 25-Mbyte cache. Intercasting uses VBI. Intel has developed a technology that allows broadcasters to inject digital data into the VBI. The company has also developed software that strips out the digital information once it reaches the PC. The signal is transmitted the same way televised signals are currently sent—either via a transmission tower, cable TV, or satellite—meaning viewers can receive Intercast broadcasts the same way they receive a regular TV signal. This enables viewing of a television program in real time in one window while simultaneously browsing through content created to supplement the broadcast. HTML pages are automatically cached on a viewer's hard drive and can be viewed during the broadcast or later. If users have a modem with a direct connection to the Internet, they can also follow hypertext links from Intercast pages to related sites on the Web.

3.10 Regulatory Environment

Although great efforts are being made for standardization, the regulatory environment for ITV is filled with uncertainty. In general, cable operators have more freedom in the ITV services market because they can offer such services under the same rules that apply to regular cable channels. Telephone companies, on the other hand, have to adhere to common carrier rules, which require them to provide all groups that wish to offer video services equal access to their network and treat all service providers equally in a "Level-1 Gateway" that users will enter when they turn on their TV sets. This is intended to prevent telephone companies from discriminating against potential competitors.

The type of content that will be provided for ITV is of great significance. There may be channels of infomercials, sports, or simply video junk that no one will want. Under current regulations, there is little that telephone companies can do to prevent this. If the gateways become cluttered with junk, consumers may be turned off or simply have difficulty locating the services they want.

The broad mix of service providers in the ITV arena—cable operators, telephone and computer companies, movie studios, and traditional broadcast networks—are also likely to lead to very different mixes of services for consumers. There are strong differences in service philosophies among these groups, as well as alternative assumptions about what consumers want. Many traditional broadcasters and cable channels see ITV as a way for consumers to have more choice and control over what they view; some computer companies perceive ITV as a video database; and many telephone companies believe that communications should be an integral component in ITV services.

3.11 Legal Issues

Many legal issues surround ITV services. For example, European directives exist regarding the number of hours per week that home shopping advertising is allowed. Another European directive states that half of a channel's programs should be produced in Europe. U.S. laws prevent telephone companies from sending video down their wires. Many laws refer to "broadcasting" but many of these services will be one-to-one (often referred to as pointcasting or monocasting).

With the convergence of publishing, common carriage, and broadcasting, many of the current laws and regulations on communications, ownership, and content will become outdated. These will all have to be rewritten or amended to accommodate changing technological scenarios.

3.12 Industry Groups

A number of industry groups are working to design complete solutions that support the video, audio, and data (V+A+D) broadcasts that make enhanced digital broadcast possible. These activities need to be coordinated to avoid duplicating efforts and fragmenting the Web by incompatible standards. The following groups are known to be working on or interested in integrating TV and Web technology: ATSC; the Association of Radio Industries and Businesses (ARIB); ATVEF; the Advanced TV Forum (ATV); the Consumer Electronics Association (CEA) [62]; the Computer and Electronics Marketing Association (CEMA) [63], which published a specification on how to send URLs over line 21 VBI; DAVIC, which developed DAVIC 1.3 DVB URL; the Digital Display Working Group (DDWG); [64] the Digital Transmission Licensing Administrator (DTLA) [65]; DVB; the European Broadcasting Union (EBU) [66]; Webcaster group IETF [67]; the URL Registration Working Group, which has defined a registration procedure for URL schemes and a set of guidelines on how to define URL-schemes; ISO MPEG-4; ISO MPEG-7 [68], a project for adding metadata to audiovisual content; (the Society of Cable Telecommunications Engineers (SCTE) [69]; the Society of Motion Picture and Television Engineers (SMPTE) [70]; and Television Anytime, an organization whose goal is "to encourage the development of TV and related multimedia services based on the use of persistent local storage irrespective of the manner of service delivery."

3.13 Summary

ITV combines aspects of traditional television and the Internet that inspires applications in e-commerce, targeted advertising, VOD, and enhanced viewing services. ITV involves moving data along with normal program content. Two data channels are typically employed in an ITV connection. One is VBI and the other is a "back channel," usually a telephone line. The VBI is typically used to send triggers and links. The back channel is used to retrieve commands for interactive content to be displayed on the television screen.

The framework to support ITV content is largely based on existing Internet standards. The objective has been to create a platform that can be supported across all television environments (analog or digital; cable, satellite, or terrestrial broadcast), and that leverages the huge base of tools, media, and know-how that has already been developed for the Web. Existing Web standards, with only minimal extensions for television integration, provide a rich set of capabilities for building enhanced TV content. In addition to leveraging the incredible investment already made in the Web itself, this framework can develop as underlying technologies grow, supporting new networks, media, and protocols. A number of standards bodies are working to develop standards.

The emergence of ITV brings about a host of exciting opportunities for broadcasters, content providers, tool developers, and equipment manufacturers. In this chapter, we looked at the various technical systems proposed or in use in ITV, developments in underlying technologies, and standards-producing bodies such as the AVTEF and DAVIC.

References

[1] Noll, A. M., *Television Technology: Fundamental and Future Prospects,* Norwood, MA: Artech House, 1988.

[2] Hartings, R. L., *Basic TV Technology,* Boston: Focal Press, 1990.

[3] Prentiss, S., *Television from Analog to Digital,* Blue Ridge Summit, PA: TAB Books, 1985.

[4] "Grand Alliance HDTV Specification 2.0," www.sarnoff.com/career_move/tech_papers/hdtv.html, September 2000.

[5] http://www.luiss.it/medialaw/lex/Europa/9238.htm, December 2000.

[6] "DVB (Digital Video Broadcast)," http://www.dvb.org/, April 2001.

[7] Saied, M., et al., "Immersive Video," *Technical Report VCL-95-108,* Visual Computing Laboratory, University of California, San Diego, California, August 1995. Submitted to VRAIS '96.

[8] "What is Hypervideo?", http://www.en.eun.org/eun.org2/eun/en/news/content.cfm?lang=en&ov=3376, December 2000.

[9] http://www.evl.uic.edu/pape/CAVE/prog/CAVEGuide.html, December 2000.

[10] Dubery, P., "Transmission in Transition," *International Broadcast Engineer, Special Issue,* 1996, pp. 26–27.

[11] http://195.195.73.7/videotxt/Videotxt.htm, April 2001.

[12] http://195.195.73.7/videotxt/teletxt.htm, April 2001.

[13] http://195.195.73.7/videotxt/viewdata.htm, April 2001.

[14] http://www.motorola.com/NSS/Products/GSM/introgsm.htm, March 2001.

[15] http://www.dectweb.com/dectforum/aboutdect/aboutdect.htm, March 2001.

[16] http://www.umts-forum.org/what_is_umts.html, March 2001.

[17] "Advanced Television Enhancement Forum," http://www.atvef.com/, April 2001.

[18] "Digital Video Broadcasting-Multimedia Home Products Protocol (DVB-MHP)," http://www.research.philips.com/pressmedia/releases/000902.html, April 2001.

[19] "Java™ Technology in Digital TV," http://java.sun.com/products/javatv/, April 2001.

[20] "The XML Cover Pages," http://www.oasis.open.org/cover.xmi.html, March 2001.

[21] "TV Anytime Forum," http://www.tv-anytime.org/, April 2001.

[22] "Liberate," http://www.liberate.com/, April 2001.

[23] "Sun Microsystems," http://www.sun.com/.

[24] "OpenTV," http://www.opentv.com/, April 2001.

[25] "Excite," http://www.excite.com/, April 2001.

[26] "ICTV," http://www.ictv.com/, April 2001.

[27] "PowerTV," http://www.powertv.com/, April 2001.

[28] "Microsoft," http://www.microsoft.com/tv/.

[29] "CableLaboratories (CableLabs)," http://www.cablelabs.com/, March 2001.

[30] "DAVIC," http://www.davic.org/, March 2001.

[31] Gall, D. L., "MPEG: A Video Compression Standard for Multimedia Applications," *Communications of the ACM,* Vol. 34, No. 1, March 1993, pp. 14–18.

[32] "ADSL Tutorial," http://www.adsl.com/adsl_tutorial.html, April 2001.

[33] "Coding of the Moving Pictures and Associated Video," *International Standard, ISO/IEC 13818,* November 1994.

[34] http://64.4.14.250/cgibin/linkrd?_lang=EN&lah=b8eac6659b071d9ff9a60b70ff068b7 b&lat=988680567&hm___action=http%3a%2f%2fwww%2eitu%2eint%2fitudoc%2 fitu %2dr%2frec%2fbt%2f601%2d5%2ehtml, April 2001.

[35] "ISO MPEG-4," http://www.cselt.it/mpeg/standards/mpeg-4/mpeg-4.html, April 2001.

[36] Done, S., "MHEG—A Multimedia Presentation Standard," http://www.doc.ic.ac.uk /~nd/surprise_96/journal/vol2/srd2/article2.html, April 2001.

[37] "Welcome to the Fascinating World of MHEG," http://www.ccett.fr/mheg/overview .htm, April 2001.

[38] "Document Markup Language HTML 4.0," http://www.w3.org/TR/REC-html40/, February 2001.

[39] Ostund, M., "Media Pool," *International Broadcast Engineer,* September 1995, pp. 24–26.

[40] http://www.halcyon.com/cliffg/uwteach/shared_info/internet_protocols.html#TCPUDP, February 2001.

[41] http://jiad.org/vol1/no1/katz/, March 2001.

[42] "IP Over VBI Internet Draft," http://www.ietf.org/internet-drafts/draft-ietf-ipvbi-nabts -03.txt-, December 2000.

[43] "rfc 2327," ftp://ftp.isi.edu/in-notes/rfc2327.txt.

[44] EIA-608, Recommended Practice for Line 21 Data Service.

[45] EIA-746A[1].

[46] "JAVA technology for ITV," http://java.sun.com/products/ javatv/, March 2001.

[47] http://java.sun.com/products/javatv/javatvspec0_9_0.html, April 2001.

[48] http://sun.systemnews.com/, April 2001.

[49] "Advanced Television Systems Committee" (ATSC), http://www.atsc.org/, April 2001.

[50] "Distributed Systems Technology Group," http://www.cmr.nist.gov/group.html, December 2001.

[51] "National Institute of Standard and Technology," http://www.nist.gov/, April 2001.

[52] "Statement of Work," http://www.nist.gov/it/div895/cmr/dase/sow.html, April 2001.

[53] "ATV Forum," http://www.avtf.org/, April 2001.

[54] "ATVEF Technical Specifications," http://www.atvef.com/specs.htm and www.atvef .com/library/spec.html, April 2001.

[55] "TV URLs," http://ds.internic.net/internet-drafts/draft-zigmond-tv-url-00.txt, December 2000.

[56] "Guide to TV Broadcast URLs," http://ietf.org/internet-drafts/draft-finseth-guide-01.txt, April 2001.

[57] "ATVEF Enhanced Content Specification," http://www.atvef.com/library/spec1_ 1a .html, May 2001.

[58] "Broadcast Markup Language," ARIB, http://lists.w3.org/Archives/Public/www-tv/ 1999OctDec/0031.html, March 2001.

[59] http://www.infowin.org/ACTS/RUS/PROJECTS/ac321.htm, December 2000.

[60] http://www.wink.com, May 2001.

[61] "Intercast Technology Brings TV to Your PC," http://coverage.cnet.com/Content/Features/Techno/Intercast/, May 2001.

[62] "CEA (Consumer Electronics Association)," http://www.ce.org/, April 2001.

---

1. Proposal for Sending URLs over EIA 608 T2, EIA-608 and EIA-746A have been produced by Consumer Electronics Manufacturers Association (CEMA) and can be purchased from Global Engineering Documents, an IHS Group Company, the premier distributor of collections of standards, specifications, and technical publications, http://global/ihs.com/.

[63] "CEMA (Computer and Electronics Marketing Association), U.S.," http://www .cema .org/, April 2001.

[64] "DDWG (Digital Display Working Group)," http://www.ddwg.org/, April 2001.

[65] "DTLA (Digital Transmission Licensing Administrator)," http://www.dtcp.org/, April 2001.

[66] "EBU (European Broadcasting Union)," http://www.ebu.ch/overview.html, May 2001.

[67] "IETF," http://www.ietf.org/, April 2001.

[68] "ISO MPEG-7," http://www.cselt.it/mpeg/standards/mpeg-7/mpeg-7.html, April 2001.

[69] "SCTE (Society of Cable Telecommunication Engineers)," http://www.scte.org/, April 2001.

[70] "SMPTE (Society of Motion Picture and Television Engineers)," http://www.smpte .org/, April 2001.

Selected Bibliography

"Broadcast HTML Standard," http://toocan.philabs.research.philips.com/misc/atsc/bhtml/, December 2000.

"Cascading Style Sheets," http://www.w3.org/TR/WD-CSS2/, December 2000.

"Content Description SDP," ftp://ftp.isi.edu/in-notes/rfc2327.txt.

"Content Encoding," ftp://ftp.isi.edu/in-notes/rfc1951.txt and ftp://ftp.isi.edu/in-notes/rfc1952.txt.

"Datagram Format IP," ftp://ftp.isi.edu/in-notes/rfc791.txt.

"Document Object Model DOM Level 0," http://www.w3.org/DOM, December 2000.

"Document Scripting Language ECMA Script," http://www.ecma.ch/stand/ecma-262.htm, October 2000.

http://www.research.microsoft.com/~mbj/papers/mitv/tr-97-18.html, October 2000.

"Hypertext Transfer Protocol (HTTP) 1.1 (rfc2068)," ftp://ftp.ietf.org/internet-drafts/draft-ietf-http-v11-spec-rev-03.txt.

"IP Delivery Over Video Broadcast (analog NTSC example)," http://ds.internic.net/internet-drafts/draft-panabaker-ip-vbi-02.txt, December 2000.

"IP Delivery Over Video Broadcast (analog PAL example)," http://ds.internic.net/internet-drafts/draft-thorne-vbi-00.txt, December 2000.

"MIME HTML (rfc2110)," ftp://ftp.isi.edu/in-notes/rfc2110.txt.

"MIME Multipart/Related (rfc2112)," ftp://ftp.isi.edu/in-notes/rfc2112.txt.

"Multicast Datagram Format Multicast IP," ftp://ftp.isi.edu/in-notes/rfc1112.txt.

"RFC 1112—Host Extensions for IP Multicasting."

"The Local Identifier (lid:) URI Scheme," http://ietf.org/internet-drafts/draft-blackketter-lid -00.txt, December 2000.

"UUIDs and GUIDs (IETF Work in Progress Draft-Leach-UUIDs-GIUDs-01)," http://www.ietf.org/internet-drafts/draft-leach-uuids-guids-01.txt, December 2000.

"SMPTE Working Group for "Standardization of Declarative Content for Television," http://lists.w3.org/Archives/Public/www-tv/2000JanMar/0017.html, March 2001.

"WebTV Networks," http://www.webtv.net/, May 2001.

4

Content Development

What is beautiful about this convergence is that barriers between audiences and creators are disappearing. This is the most exciting development of our time.

—Meryl Marshall, CEO of the Academy of
Television Arts and Sciences

4.1 Introduction

TV broadcast concerns both digital and analog systems and includes systems like DVB, ATSC, DSS, NTSC, and PAL. The TV broadcast "network layer" is non-IP based. Typically, TV broadcast systems are push systems. The service provider schedules the broadcast of content via a TV transport system, and the user has no influence on it. In this model, the user accesses the stream(s) rather than the server at the upstream station.

ITV is basically domestic television boosted by interactive functions generally supplied through a back channel or modern terminal. The distinguishing feature of ITV is the possibility that new digital technologies give the user the ability to interact with the content offered. Limited interactivity takes place in the analog system by sending data in the vertical blanking interval (VBI).

TV broadcast content is modeled in a four-layer hierarchy, consisting of service, event, component, and fragment (Figure 4.1). Service is at the top, while fragment is at the bottom of the hierarchy. The term *service* is used to

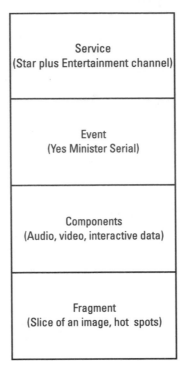

Figure 4.1 TV broadcast content hierarchy.

refer to a concatenation of programs, all being broadcast by the same service provider. The programs of a service share some tuning characteristics. Service also corresponds to the naming of the channel, such as "Entertainment Channel" as used in today's analog TV.

The term *event* is used to refer to a single TV program. An event consumes a time period within a service and, therefore, can be characterized with begin and end times. The service provider determines the granularity in which the service is split in events. An event can be a complete program or an episode of a program. Events can be grouped in series, for example, to form a serial. Events are the typical entities that EPGs list to present program schedule information.

The term *component* is used to refer to the constituents of an event. Examples of components are the audio and video of a TV program. Multiple audio components exist for multilingual programs. Components for interactive programs are application documents and other data, such as the graphics or multimedia that these applications use. In addition to continuous data like audio and video, components encompass discrete data like Web pages and applications describing composition and interactivity. The term *fragment* is used to refer to a subpart of a component. For instance, it can be a slice of a video sequence, a subregion in an image, or a hot spot.

Due to the push character of TV broadcast, there are two dimensions of hierarchy: a schedule-related hierarchy and a content-related hierarchy. The first is the hierarchy of transport system, transport stream, service, series, and event. The second is the hierarchy of series, event, component, and fragment.

At a very high level, the basic stages required to make and broadcast an interactive program are the same for all of the constituent components, be it video, audio, or presentation data. The four stages are capture, edit, play, and broadcast. The availability of mature production tools and techniques has simplified the handling of video and audio components. Both are generally captured real time and at the same time, although the increasing use and sophistication of post-production sound and vision manipulation is making this less true. Also, they are generally stored on the same physical media beginning at point of capture. This greatly simplifies the task of maintaining synchronization in downstream stages.

Editing video and audio only requires temporal coordination. This is because separate presentation devices are used—typically a screen for video and loudspeakers for audio. Playback of video and audio components in DTV systems is based on the mature standards for MPEG encoding, compression, and transport.

When adding a data component to provide a text presentation, there is no coordination either temporally or spatially with video or audio. In most occasions, integrating the data only needs to occur at the time of broadcast. The video and audio may have been captured on location months ago, edited weeks ago, and loaded into a playback system days ago. Meanwhile, the text presentation could be a display of an image related to a breaking news story, captured, edited, and loaded into the playback system only moments ago. A number of standardization initiatives are looking at how to handle such metadata at all stages of production [1].

Data is delivered to the receiver using a broadcast carousel or on-line connection (Figure 4.2). The principle of a data carousel is that modules of data are transmitted in a cyclic manner. A receiver simply waits for the data it requires to be retransmitted. In both cases, there is uncertainty in the time taken to deliver an object into the receiver. Clearly, preloading data into receivers well in advance and then sending only simple synchronizing triggers is an obvious solution, but this may not always be possible.

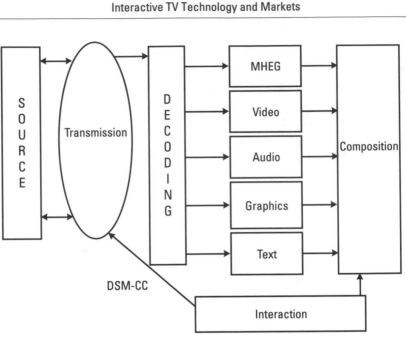

Figure 4.2 Functioning of ITV.

In this chapter, we will first examine the types of enhancement required in ITV content and then discuss how to accomplishing them. We will also look at a specific example of WebTV and see various aspects of its programming.

4.2   Content: In Search of the Killer Application

Content plays the most important role in making ITV popular. There has been much discussion about potential killer applications—services or content that will be almost irresistible to consumers and will drive ITV into millions of homes. Historically, some media have benefited from such killer applications while others have required a critical mass of titles or applications. E-mail was a killer application for the Internet, while CD players and VCRs needed a critical mass for their success. Movies-on-demand and interactive games could be the two killer applications for ITV. However, ITV will need a critical mass of households to succeed. Toward this end, attractive and imaginative content will play a crucial role.

Historically, new media is filled with content from earlier media. Early radio content came from vaudeville acts, and early television content came from radio shows converted to a television format. Today, virtually every proposed service for ITV is a modified version of content or a service that already exists. Movies-on-demand is really a videocassette rental service directly to the home. Most of the proposed interactive home shopping channels are variations of QVC and Home Shopping Network or video renditions of catalog shopping. Interactive games allow two people in different locations to play a game together without physically meeting each other. Although these enhancements to existing services are attractive, they are not bold, new applications for a new medium. New applications as a result of creative innovations have to appear for ITV to succeed.

Original interactive programming will be expensive. In a start-up phase, program providers are reluctant to invest heavily in original content that attracts only a relatively small number of viewers. At the same time, some programming, such as news, sports, and game shows, can be adapted to interactive formats with only modest increases in production costs. However, original content and new program formats are required to fully exploit ITV's potential. ITV content may be a mix of the two, such as enhancing existing content and offering completely new, innovative formats. However, one thing is certain. Content creation for ITV is much more than integrating TV into the Web. Let us first discuss what viewers want and what ITV provides.

4.2.1   What Do Viewers Want?

The customer is the king. Programming's success depends on viewers' acceptance of its content and format. Viewers do not want the broad array of Web content they can call up on their PCs. TV is an entertainment medium, and content providers should keep that in mind when introducing new interactive services. A concept called lazy interactivity involves being able to click to view something with a remote control in one hand and a beer in the other. TV was made for pure entertainment. TV is a spontaneous activity, where people are enjoying their leisure time, while the PC is a more deliberate medium.

The "mouse potato" culture will have to occur in the living room, where Dad may have to wait until *Who Wants to be a Millionaire?* hits a commercial break, at least if he wants to preserve peace in the family [2]. The interactivity has to be simple and relaxed and much different from what is available on the PC.

Survey data shows that viewers with access to enhanced TV content watch up to one-third more television programming than before. At a time when viewers get increasingly distracted by the Internet, prerecorded videos, and video games, broadcasters and postproduction houses need every audience edge they can get. Content creation for enhanced TV is, therefore, an involved job.

A series of ITV companies (ACTV, Liberty Livewire, Motorola, OpenTV, and Universal Electronics, a handheld device maker) hired Boyd Consulting to do market research to discover American perceptions about ITV. About 500 participants (head of households) were asked a few one-on-one questions. The questions covered whether they would prefer a two-screen experience (synchronized TV), a single-screen experience (enhanced TV), or if they would like a handheld unit to interact with their television. Survey results indicate that the American public is increasingly more aware of what ITV is and can offer [3]. The following are some of the results:

- 66% preferred enhanced TV supported by the handheld device;

- 76% liked a handheld to support the TV experience;

- 52% wanted news, weather, and sports information from ITV;

- 44% wanted a TV guide;

- 38% wanted behind-the-scenes information on programs;

- 37% wanted e-mail on TV;

- 32% wanted games or quizzes;

- 47% analog customers said ITV would make them upgrade;

- 57% said if the enhanced TV service had a handheld as well, they would upgrade;

- 42% of satellite customers said they would switch to cable if these services were only available on cable.

Viewers distinctly gave their preference for the handheld remote. This means the interactivity has to be quick and simple. Further, news, weather, and sports are the preferred programs. This may simply be because viewers currently receive these programs and they could not think of other types of programs they would like to receive. Content creators have to keep in mind the above results of user preferences.

4.2.2 Commercials

Commercials are crucial to developing and sustaining ITV. One of ITV's strengths is its ability to segment audiences. It can provide different content to different audience segments and know which audience segments are watching. It can provide instantaneous feedback from audiences, including the direct purchase of products and services. These characteristics need different business models for advertisements on ITV. Most likely a hybrid model will emerge in which ITV serves as a traditional mass medium for some applications and a telemarketing medium for others.

Consumers' response to ITV commercials is also important. Consumers can choose which commercial or which version of a commercial they watch. They can then request and receive additional information or actually make a purchase during a commercial. Advertisers could also run contests in which a viewer might instantly win a prize. These features may make commercials more appealing to viewers and potentially reduce the incidence of channel changing during spots. There will undoubtedly be a need for a novelty effect with these new forms of commercials. Therefore, the commercial has to be produced in quite a different way to serve ITV viewers.

Interactive commercials may serve some existing needs, such as devising ways to reduce channel changing. ITV can be used to provide new settings in which advertisers can develop new formats. These formats cannot be introduced in a traditional setting. For example, the development of specialized cable channels such as CNBC supported an environment in which it was possible to place program content and a commercial message on the same screen. This type of technique was unheard of in a broadcast network program. In established traditional programming, content cannot be mixed with commercials.

Many innovative commercials have also been tried. British Airways developed a novel interactive commercial for theaters in the United Kingdom to sell its flight tickets: On screen, a couple walks through a park in Paris. Suddenly, a woman in the theater shouts, "That's my boyfriend. That's Nigel." The on-screen boyfriend turns to the camera and says, "Michelle, I can explain everything." The interaction then continues between the boyfriend on screen and the woman who was planted in the theater. This novel and creative commercial requires considerable logistical support in a theater environment. However, the concept lends itself quite well to an ITV environment that speaks to narrow audience segments (in the near term) and potentially to each individual viewer (in the long term). The commercial provides an example of a radical departure from the norm.

Many technological and behavioral changes are transforming television, such as the development of megachannel cable TV systems, menu-driven user interfaces for TV (in which viewers can navigate through large-capacity cable TV systems), and increased channel changing by viewers. The design and placement of commercials has been affected by these trends. In an effort to reduce channel changing, many stations and networks adopt "hot switching" or "seamless programming," where the commercial gets integrated with the main program.

The development of interactive commercials will take place in the context of all these technological and behavioral changes. Radical new forms of advertising are emerging in an ITV environment.

4.2.3 Walled Gardens

The walled garden is a controlled environment where merchant sites are made accessible to viewers for e-business. Walled gardens allow network operators, e-businesses, and advertisers to create an interactive experience with the emotion that TV can deliver, coupled with the immediate gratification of information access and electronic purchasing that the Internet can deliver. Walled gardens are secure environments, with a common TV-centric user interface, personalization capabilities, and cross-merchant shopping capabilities. The content in the walled garden is optimized for TV viewing and managed by the network operator. Home shopping, home banking, various news services, and weather or sports information providers provide information. E-commerce operators provide facilities for electronic transactions, including credit card authentication. Content and merchant sites are designed for a TV environment and linked to the TV portal platform through the walled garden. The following are the key components of a walled garden:

Personalized Information   These are on-demand information services available to end users based on their preferences or profile, including headline news, sports news and scores, local weather, business news, horoscopes, and stock quotes. For personalized information, user preferences are captured and stored. When users log in to access their part of the walled garden, based on the preferences stored in a database, the information specific to their choices is provided.

Branded Information   These are information services in various categories from branded sources.

Shopping These are facilities for "t-commerce" in categories such as books, video, music, gifts, electronics, apparel, and toys. Advanced shopping features such as comparison shopping, cross-merchant shopping carts, and auctions are also provided.

Advertising Advertising opportunities are available in the walled garden similar to Internet portals (banner ads, etc.), but also to leverage the interactive, media-rich capabilities of the TV.

4.2.4 Extended TV Services

Interactivity takes everyday TV functions and extends them into new value-added services. Extended TV services include interactive EPGs, personal video recording capabilities, VOD, e-mail access, chat functions, and more. These services can all be accessed through the network operator's TV portal. The EPG is the navigation tool for TV viewing. It is an essential tool for ITV users, especially as the number of channels and other offerings increase. Personal Video Recorders (PVR) store video programming on an internal hard drive and function like a personalized video server, allowing end users to "time-shift" their TV viewing. PVRs also act as autonomous agents, searching all of the available programming and building a dynamic menu of personalized content choices that users can access at their convenience. VOD allows users to play, pause, and rewind videos on their TVs via their remote control. E-mail access, chat functions, music jukeboxes, and photo albums are examples of other extended TV offerings.

All of the above TV services and applications need to be incorporated into the portal and presented to end users with a common user interface and order/transaction entry and tracking system. Each application has specific integration issues that may arise, depending on the environment that network operators choose to deploy. Integrating into the base portal platform alleviates many of these issues and allows for a more fluid experience within the overall portal.

4.3 Creating an ITV Program

ITV is the result of a convergence between television and the Internet. Web sites can take advantage of broadcast video by inserting broadcast signals into their pages, while broadcasters can use Web sites to complement the viewing experience of popular shows. The first step in creating an ITV program is to create Web pages designed specifically for a particular TV show. It is a much better idea to design Web pages that complement what is happening in a TV show than to simply link viewers to a broadcaster's Web site, because unrelated Web pages can be distracting and can even lead viewers away from the Web site. When Web content fits the program well, viewers will continue to watch the TV show and explore the interactive program at the same time. For example, broadcasters can create Web pages that contain cricket players' statistics. During the game, broadcasters can display links that point to information on the player currently batting. The viewer can watch the game and read the player's statistics at the same time. The other important factor is related to design issues when Web pages are being developed for TV viewing. We will examine these issues in further detail later in this chapter.

4.3.1   ITV Links

TV broadcast applications need a mechanism to identify and locate the components building the application. ITV links are links between TV broadcast and Web content. These links open possibilities for seamless transition in referencing resources at TV broadcast transport and Internet sites. ITV links provide an easy and powerful way to bring the richness and interactive nature of the Web to the television audience. ITV links can be used to show television viewers Web content specifically designed to complement what they are watching on TV at any given time. To enhance the viewing experience even more, ITV links can be used to create truly interactive programming by integrating the content of the Web pages with the television broadcast itself. These links are broadcast within the TV program, in a portion of an analog TV signal called the VBI. They are transmitted as part of the television broadcast and work in conjunction with HTML tags in associated Web pages. URIs are much the same as the more familiar Web-based URLs but with a broader definition. URIs point to the HTML pages and related resources that comprise the interactive portion of the programming.

4.3.1.1   Global and Local URIs

Global refers to URIs that can be accessed anywhere around the world and that identify content related to any TV broadcast system in the world, including storage devices associated with that system. Such a global URI may include identifying the TV broadcast system to be used. *Local* refers to URIs that are accessed within a certain TV broadcast system and that identify content to be accessed through that system. URIs that refer content outside the local TV broadcast system are assumed to be either global URIs or traditional

URIs for locating resources on the Internet. Identification of a particular content item belongs to the local class of URIs. Global URIs typically refer to a service or an event. However, an exception can be found in the case where the same content item is referenced in various transport contexts, such as in a commercial.

As a set of resources and their various locations can scale to large numbers, it is preferred that the URI scheme imposes a hierarchical structure when the URI's purpose is to locate a resource. A hierarchy allows for step-by-step resolution and navigation to the resource identified. Further, a hierarchical structure allows one to group resources. It also helps differentiate between various components, for example, between a serial and an event in that serial.

Application Scenario

URIs can be used to enhance TV programs in several ways. A few of the possible scenarios are described below:

Locating EPG   URIs are used for referencing EPGs, giving TV broadcast services and events from a Web page to navigate to them. The references allow modifications in the actual transmission schedule, but a coarse indication can be derived. The broadcast program can be indicated through tuning data or naming. The EPG also supports setting reminders or recording programs. It is also possible to record all episodes of a serial.

Referencing Audio-Video of a Broadcast Via Web   A Web page is composed for presentation on a TV broadcast receiver. The Web page is delivered in association with a TV broadcast program. The transmission paths may be physically separated. The Web page includes an object, which refers to the associated audio-video image of the TV broadcast program. URIs help in referencing the audio-video.

Referencing to Unknown Location   In a Web page, a TV broadcast event is referred, but the exact location is not known at authoring time. The URL is kept incomplete in its information. A query is added to retrieve the missing information. When the available TV broadcast system supports the query mechanism, the URL is resolved and the identified resource is retrieved. The query language is technology-independent. It does not rely on specific fields, such as SI data, in the TV broadcast transmission system. Examples are `dtv://?program=X-Files` and `dtv://TVI/?lang=en`.

Data Enhancement During a Program A data file can be transmitted along with a TV program. The data file may contain additional information to that program. It may also contain hyperlinks to the programs and data in other data files being broadcast on the same channel and in other channels, so that receivers can set reminders for the upcoming game or data file.

A TV broadcast of a cricket match can be data-enhanced. This is done by repeatedly sending a file containing up-to-the-second statistics on runs scored, wickets out, bowling statistics, and other information, in a data carousel module or an encapsulated IP datagram. The broadcaster puts a URL on the Web site that references this file. Internet-connected TV receivers all over the world can get this file and display it.

Enhancing Commercials A Web page is transmitted with a TV broadcast commercial. The commercial is about an upcoming TV broadcast program. The viewer can click a hot spot area to set a reminder for that program. The Web page can also be accessed at the broadcaster's Web site.

A set of three Web pages is transmitted with a TV broadcast commercial. The viewer can navigate the three pages. The pages are transmitted frequently along various TV broadcast systems. The pages can also be accessed at the advertiser's Web site, where they are maintained at a particular subdirectory. The advertiser uses relative referencing between the pages.

Enhancing Quiz Show A live quiz show is enhanced so that the viewer can play along. The enhancement data are a mixture of Web pages, which compose the quiz's question-and-answer environment. These include element values, which carry the actual questions and (correct) answers to be inserted in the Web pages, and procedural cells to control the viewer's score. The Web pages are provided at a Web site long before the show is broadcast, so that viewers can prepare. The element values are transported along the TV broadcast transmission channel during the show. They are synchronized with the actions in the show so as to complete and update the application.

There are several levels of play. For example, some pages may provide the viewer with hints to ease answering. Some pages provide fewer alternatives in the multiple-choice questions. The viewer can select the level by navigating between these pages. On the day of the broadcast, the content is enhanced. The enhancements are also broadcast along with the program. The enhancement refers to the Web site so that, upon tuning into the TV broadcast, the Web site's home page gets retrieved. Triggered by stream events in the TV broadcast transport stream, the enhancement also controls the insertion of element values (questions and answers) and score management (for example, no scores can be entered after the answer is given).

4.3.1.2   How ITV Links Work

ITV links contain text codes called URLs. These point to specific content —Web pages—on the World Wide Web. ITV links are transmitted as part of the television broadcast in the same manner as closed captioning text. Web pages are created using HTML which is composed of short strings of symbols and text, known as tags, that are placed around the content that viewers see. These tags tell receivers how to display the content contained in the Web pages. When a TV viewer chooses to display the interactive show, the receiver interprets the code in the ITV link and displays the appropriate Web page. ITV receivers are required to support the following:

- MIME-type text/HTML 4.0 (Frameset Document Type Definition);
- Text/CSS 1;
- ECMAScript;
- DOM 0 (JavaScript 1.1);
- Image/png (no progressive encoding);
- Image/jpg (no progressive encoding);
- Audio/basic.

Next to locating components at TV broadcast transport channels, the links are used to reference events. Events are accessible at the TV broadcast transport channel, possibly at several channels and at multiple periods of time. The content may be stored and made available through another path than the TV broadcast transport channel. Most evident are local storage, VCR-type devices, and the Internet itself. Local storage devices can be connected through an in-home network to the user agent presenting the application.

TV broadcast content delivered through an IP tunnel is considered as content is made available through the Internet. An IP tunnel refers to a forward path, which is logically separated from the conventional TV broadcast transport protocol but uses the same physical transmission link.

HTML Extensions for ITV

Because ITV involves displaying Web content and television simultaneously, the receiver must determine how to display the information. Web pages are created using HTML, which describes how to handle the information between the tags. Two new HTML extensions have been developed for ITV display. One tells the receiver if it should display information as TV or as Web content. The other defines the package of files that make up the ITV program and tells the receiver where to find it.

4.3.1.3 Creating ITV Links

ITV links are straightforward and easy to create. They consist of a URL and any number of attributes, depending on the interactive show's design and needs. Abbreviations are available for all of the attributes, except URL and checksum. ITV links with abbreviated attributes are transmitted more quickly and more accurately.

ITV Link Syntax

The ITV link-encoding format is specified by the Electronic Industries Association specification EIA-746-A, "Transport of Internet Uniform Resource Locator (URL) Information Using Text-2 (T-2) Service" [4]. This specification is available on-line for $33 on the Global Engineering Documents Web site [5, 6]. For United States and Canadian broadcasts, ITV links are transmitted on line 21 of the VBI of an analog television signal. Line 21 is the part of the VBI that is also used to send closed captioning text. The standard accepted by the Computer Electronics Manufacturers Association (CEMA) mandates that the ITV link string is inserted in Text-2. We discussed global and local URIs earlier. We will discuss the syntax for the links for clarity.

Size Limitations The total number of characters that can be transmitted in an ITV link is limited by the physical constraints of the VBI (2 characters/frame and 30 frames/second) and by the encoder. The encoder determines how many characters per second it can encode, and therefore how long the ITV links can be. The technical specifications for transmission characteristics are as follows:

> *Characters per second:* 60 characters can be transmitted per second.

> *String length:* An average string is between 45 and 60 characters.

> *Bandwidth:* EIA-746 recommends that ITV links are limited to 15% of the available bandwidth on field 1 of line 21. However, when the closed captioning is encoded before or during ITV link encoding, it is generally safe to use all of the remaining field 1 bandwidth for ITV links.

Triggers   Triggers are real-time events delivered for the enhanced TV program (see Sections 3.5 and 3.7.5.1). Receivers use trigger arrival as a signal to notify users of enhanced content availability. Triggers always include a URL and may optionally also include a human-readable name, an expiration date, and a script. Triggers that do not include a `name` attribute are not intended to initiate an enhancement and are processed as events, which affect (through the `script` attribute) enhancements that are currently active. If the URL matches the current top-level page, and the expiration has not been reached, the script is executed on that page through the trigger receiver object. Triggers are text based, and their syntax follows the basic format of the EIA-746A standard.

ITV Link Attributes

ITV links can have any combination of a number of attributes. The attributes are `url`, `type`, `name`, `script`, `expires`, `view`, `checksum`. The general format for triggers (consistent with EIA-746A) is a required URL followed by zero or more attribute-value pairs and an optional `checksum`:

```
<url>  [attr1: val1][attr2:val2]...[attrn:valn][checksum]
```

All of the attributes, except `URL` and `checksum`, can and should be abbreviated to increase the speed and accuracy with which they are transmitted. `URL` and `checksum` are the only required fields. The detailed attributes, along with their syntax, are given in Appendix 4.A. Illustrative attributes for the trigger are given in Table 4.1.

Table 4.1
Illustrative Attribute for Trigger

```
<http://air.kode.net/main.html>

<http://air.kode.net/main.html[name:Entertainment
Channel!]>

<lid//air.kode.net/main.html[n:Entertainment Channel!]>

<lid//air.kode.net/main.html[n:Entertainment Channel!]
[e:20011231T115959][s:frame1.src="http://dd.com/fra-
me1"]>

<http://www.newmfr.com[name:New][C015]>
```

The view *Attribute*

HTML is collection of tags (see Section 4.3.1.2). Certain parameters are used to modify a tag. For example, the [font] tag can have a color and size parameters associated with it to specify the size and color of the text.

```
<font size=+2 color=blue This Is Large Blue Text </font>
```

An anchor tag, or other selectable item, may contain a view=web parameter. This lets the receiver know that the resource referenced by the link should be displayed as a Web page.

For example, if the link <href= http://air.kode.net/ view=web> is followed from a TV program by selecting the ITV link icon, the display will change from TV to Web mode. If the view parameter is not included, the view remains unchanged. The view parameter is used only on pages that are already displaying TV.

ITV Link Checksum *Attribute*

Checksum is a means of verifying if all of the information (data bits) contained in a link is retrieved by a receiver. A checksum is a total count of the number of data bits contained in a transmission. This number is included as part of the transmission so the receiver can verify that the correct number of bits arrived. When the numbers match, it is assumed that the transmission was successful.

The Local Identifier URL Scheme: lid:

Content delivered by a one-way broadcast is not necessarily available on demand. It is in contrast to when HTTP or FTP delivers content. For such content, it is necessary to have a local name for each resource. To support cross-references within the content (for use in hyperlinks or to embed one piece of content in another), these local names must be location-independent. The lid: URL scheme enables content creators to assign unique identifiers to each resource relative to a given namespace. The syntax of the lid: URL is as follows:

```
lid://{namespace-id}[/{resource-path}]
```

Transport *Attribute*

In addition to the other attributes used in triggers, ATVEF Transport Type A triggers must contain an additional attribute, tve:. The tve: attribute

indicates to the receiver that the content described in the trigger is in conformance to the ATVEF content-specification level. For example, [tve:1.0]
Transport Type B uses the same syntax for triggers as Type A.

CDF Files and the Link Tag

ITV programs usually consist of many Web pages and graphics. They can be
listed in a single file so the receiver can preload pages and store them locally
when content is delivered via a direct Internet connection. These files are
defined by the Channel Definition Format (CDF) [7] and have the file
extension .cdf. The CDF file that contains the ITV content is referenced via
an HTML tag on the main content Web page. Example:

```
<LINK rel=package src= http://www.airkode.net/superpkg.cdf
```

CDF Example

In the above example, "http://www.airkode.net/superpkg.cdf" is the URL
address of the ITV link. When the receiver sees this URL, it retrieves the
"superpkg.cdf" file and reads it. It then downloads everything listed in that
CDF file (including all dependent resources, like images) and stores them
locally (often in a cache). For example:

```
<?XML Version= 1.0  Encoding= iso-8859-1 ?>
<Channel HREF= overlay.html  BASE= http://www.air-
kode.net/  SELF= channel.cdf

<!Each<Item> designates a part of the TV applica-
tion. The client will use this information to cache
the complete TV application. >

<Item HREF= http://www.airkode.net/itv/de-
mo1/re_overlay_01.htm >
<A HREF= http://www.airkode.net/itvl/demo1/re_over-
lay_01.htm ></A>
<Title>TV Overlay Demo PageTitle</Title>
<Item>

<Item HREF= http://www.airkode.net/itv/demo1/re_em-
bed_01.htm >
 <A HREF= http://www.airkode.net/itv/demo1/re_em-
```

```
bed_01.htm ></A>
 <Title>Embedded TV Demo PageTitle</Title>
<Item>

</Channel>
```

4.3.2   Displaying TV Broadcasts and Picture in the Web Page

It is easy to create compelling ITV shows by integrating live TV broadcasts in
the Web pages. To do this, HTML tags that tell the receiver how to display
the TV are used. The result is seamless, fully integrated interactive programming. (For the examples given in this section, HTML 4.0 syntax will be
used.)

There are two main components of an ITV show. The first is the interactive TV link that is broadcast with the TV program. The second is the
Web content specified by those links. When ITV content originates from the
Web, it is essential to make sure the video is displayed properly in the Web
page. To do this, HTML extensions designed specifically for interactive TV
are used to define how and where the video should appear within the Web
page.

After navigating to the Web page specified in the ITV link, the receiver
needs to know that a specific embedded object within an HTML page
should be displayed as video. The tv: URL is used for this purpose. The
tv: protocol places the TV picture on a Web page. It is also used to display a
TV channel without any Web page enhancement. For example, inclusion of
HREF=tv: in the Web page results in tuning of the current channel and display of channel as full-screen television. Generally, it is best to simply use
tv: to tune back to full-screen video (from a Web page with a reduced-size
TV display) without specifying a channel. This is because viewers may be
watching the show on videotape or on channels other than the one specified
in the tag (for instance, on cable). Within the tv: attribute, TV stations can
be referred by channel number, network, or broadcast call letters. For example, the tag tv:12 indicates the current broadcast on tuner channel 12,
where tv:xyz indicates that the tuner should find the current broadcast by
the XYZ network, and tv:xzzy should tune to the broadcast by the station
XZZY. The tag tv: indicates that broadcast video should be displayed where
specified without changing the current tuning settings. In the following
example, it tunes to channel 12.

```
<img width= 100   height= 75   src= tv:12>
```

Depending on what kinds of broadcast information are available to the system, or what tuning tables are used, it is also possible to identify a channel by name or network. Names are not case sensitive.

```
<img width= 100   height= 75   src= tv:WDAQ>
<img width= 100   height= 75   src= tv:abc>
```

TV Navigator and DTV Navigator allow use of the TV signal just like an <IMG> tag. Using the tv: protocol to put TV on a Web page, it can be assigned to a single image, because there is only one tuner. To add TV as a foreground image on a page, tv: as the src attribute of the <img> tag is assigned, such as:

```
<img width= 200   height= 150   name= tvpicture
align= right
src= tv:abc >
```

To embed broadcast television content in a Web page, the tv: attribute, in conjunction with other common HTML tags such as <img>, <object>, and <backround>, is used. The effect is a Web page that fills the screen with a video object of the specified size embedded in it.

```
<img src= tv:   width=320 height=240>
<img   src=   tv:12   height=90 width=120>
<object data= tv:   width= 50%   height= 50% >
<object data=   tv:XYYZ   height=90 width=120>
```

To put TV as a full-screen background of the Web page, the background attribute of the tag is used.

```
<body Background= TV:
<body style= background: url(tv:)
```

The td tag is used to place the TV picture as the background of a table cell:

```
<td width=320 height=240 style= background:
url(tv:)
```

Content overlaid on top of the TV picture inside this table cell.

```
</td>
```

When full-screen TV is desired beneath the frames, a frameset style tag is used:

```
<frameset style= background: url(tv:)   cols= 200,*>
```

TV images can also be layered on top of Web pages, using z-ordered (three-dimensional) cascading style sheets. In this case, the video should be set to Z=-1 to make it the bottom layer.

```
<div style= POSITION: ABSOLUTE; Z-INDEX:-1
<object data= tv:   height=100% width=100%
</object>
</div >
```

For television to be viewable behind Web content, transparency features are used. Placing text directly over a TV signal risks making the text difficult or impossible to read. When text is displayed on top of a television picture, a semitransparent background behind the text is provided. In this case, a frameset with TV as a background image is used. The subframes are transparent, with the bottom frame including a background image. Each frame in the frameset that wants the full-screen TV to show through must specify a transparent background color in the body tag of the frame's HTML document:

```
<body style= background: transparent >
```

Any transparent HTML components, such as table cells or images, allow the television picture to show through (Table 4.2). A screen shot showing transparent text over a background image is shown in Figure 4.3.

To prevent distortion, a container for the TV picture proportional to the TV screen is used. Otherwise, the picture will be stretched or squashed to make it fit. In North America, NTSC video format with a 4:3 aspect ratio, width to height, is used. In Europe and other parts of the world, PAL video format with a 5:4 aspect ratio is used.

Table 4.2
Use of Transparency

```
<TABLE BGCOLOR= EFEFEF   BORDER=2 cellpadding=2
<TR>
<TD>Monday </TD>
<TD BGCOLOR= Transparent ><FONT COLOR= EFEFEF Tues-
day</FONT></TD>
<TC>Wednesday</TD>
</TR>
<TR BGCOLOR= Transparent>
<TD COLOR= EFEFEF >Thursday</FONT></TD>
<TD BGCOLOR= EFEFEF >Friday</TD>
<TD><FONT COLOR= EFEFEF >Saturday</FONT></TD>
<TD BGCOLOR= EFEFEF >Sunday</TD>
</TR>
<TR>
<TD COLSPAN= 4 ><A HREF= 04b.html   target= middle >View
Source</A></TD>
</TR>
</TABLE>
```

The use of `tv:` as the href of an anchor tag allows for hyperlinking to full-screen TV:

```
<a href= tv:> Click here to return to TV</a>
```

TV Safe Area

The total scanned area of an NTSC monitor is 640 pixels wide by 480 pixels high, but a typical receiver does not show all of that to the viewer. This is because the screen is over-scanned by a certain amount to ensure that the edges of the scanned area never show to the viewer. A "safe area" (556 pixels wide by 416 pixels high) has been defined for the amount of the total scanned area that viewers will always be able to see. Around this area, a buffer zone fills the remaining screen space. The default color of this buffer zone is the same as the background color of the Web page. The background image for a Web page must stay in the TV safe area, but the background color may fill the entire screen. This allows a designer to integrate graphics and text together to create a layered effect. It is important to keep background images

Figure 4.3 Screen showing background against transparent text.

down to a reasonable size (less than 50K). Using this kind of layered effect requires some delicate design technique, requiring precise layout. If the text layer does not line up with the background, it may look strange.

4.3.3   Companies Creating ITV Links

Lists of companies that produce, develop, or deal with ITV programming and provide the hard- and software are given in Appendix 4.B. This list is not exhaustive.

4.4   Design Considerations

Designing content for enhanced TV needs careful consideration of several aspects, since pages will be shown to viewers on a TV screen, not a PC monitor. A TV screen has a lower resolution than a modern computer monitor. Its width is about 556 pixels for NTSC and about 720 pixels for PAL. This is against a screen resolution of 1024 pixels for a computer screen (see Section 3.3.6.6.). The display on a TV screen is poor, especially for colors. The display is worse if the Web page information comes in through the NTSC or PAL modulation and best if the interface from the Web page source is RGB baseband video. However, most pages look fine on TV without any extra work on the part of the designer because TV Navigator automatically resizes images, wraps long lines of text, tweaks text for better TV display, increases the default font size, and so on. However, it is a good idea to design pages to fit in the first place. Web pages contain a lot of static text and artwork, which do not display as crisply on a TV as on a PC monitor. Users view TV from a distance. If too many details are placed in a small area of the screen, viewers will find it difficult to read. As such, while designing content for enhanced TV, text should be relatively large, simple, and well spaced.

TV Navigator increases the size of fonts to 18 points (default font size), compared with 12 points for computer-oriented browsers. The distance between the user and the TV screen, as well as the fuzzier nature of a TV display, require this adjustment. However, the larger font sizes mean that less text will fit on a TV screen than in a comparable area of a PC monitor. For example, a table cell that is 100 pixels wide will hold fewer characters when displayed on TV than when displayed on a computer screen. Moreover, the automatic increase in font size does not affect text in images; for example, a GIF image used as a button label or as a link. If images that contain text are to be used, the designer has to ensure that the text is in a font large enough for television.

4.4.1 Color

Colors display differently on a TV screen than on a computer monitor. The difference is most pronounced with vivid colors. They look hotter on a TV than on a computer. TV Navigator automatically modifies some colors to make them look better on TV. However, it is better to design pages so those color problems do not arise at all. Some TV sets generate visual artifacts such as chroma crawl, bleeding, bloom, moire patterns, and flicker that do not appear on computer monitors. These artifacts can be avoided by choosing the right color combinations and staying away from certain patterns.

Chroma crawl occurs where two very different colors are placed next to each other. If the color (or chrominance) difference between the two is large, letters against a solid background appear to jiggle. Images and text look fuzzy. Reducing the chrominance distance between adjacent colors can minimize chroma crawl. One way to do that is to make them less sharp, less dense, and less vibrant. For TV, it is better to use pastels rather than intense or vivid colors. High red, green, and blue component values are not used.

RGB values are kept somewhere under 240, instead of the maximum 255. Intensely colored text on a black background or black text on an intensely colored background is not to be used. Other techniques used are to blur out the edges or add a drop shadow.

Colors sometimes bleed into surrounding areas. Bleeding often accompanies chroma crawl, but it also occurs by itself. It can be fixed the same way that chroma crawl is fixed by reducing the richness of colors and sharp chrominance transitions.

4.4.2   Bloom

Many TV sets bloom. On TV, the width of a scan line can vary depending on the brightness of the content of that line. Let us consider the case of a brightly colored box with bright text against a dark background. The scan lines that travel through the text will not be of the same length as those that travel through the darker areas between the lines of text. Therefore, the vertical sides of the box will bulge outward or bend inward depending on where the text is. To prevent bloom, highly contrast colors should be avoided.

4.4.3   Moire

Moire patterns are patterns that superimpose themselves on other patterns. Figure 4.4 shows the kind of pattern that would cause moire. For example, a gray herringbone jacket can generate a rainbow of color on TV. Red can bleed from a series of closely spaced white and black lines. The only way to prevent moire effects is to avoid the detailed patterns that generate them. Dot patterns and narrow, contrasting lines that are tightly spaced should not be used.

4.4.4   Flicker

Flicker occurs because TV monitors are interlaced. A thin horizontal line just one-pixel thick appears to flicker on and off. Similarly, horizontal edges between two contrasting colors pulsate. A flicker filter minimizes these effects. TV Navigator includes a software flicker filter if the hardware does not have one. One result of the filter is that horizontal lines and edges often appear to have lower contrast on a TV screen than on a VGA display. Therefore, it is a good idea to avoid very narrow horizontal lines and horizontal edges between contrast colors.

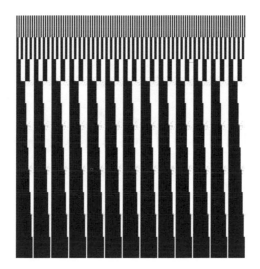

Figure 4.4 A pattern that would cause moire on a TV screen.

4.5 Enhanced TV Coding

A few sample codes for enhanced TV are given in this section to clarify the process. These are simplified examples and do not show all the code that would normally be used. The codes use ATVEF [8] standards. Table 4.3 shows a code for displaying a sample message.

Much of the functionality that is required by a television content developer is provided by the Java platform on the receiver (see Section 3.6). Java TV API is an extension to the Java platform for interactive TV. Java TV API provides access to the service information database, service selection, TV-specific media player control, and data that is broadcast with an analog or digital television signal. Using the Java TV API, developers can enable the functionality in many types of television-specific applications, such as EPGs, program-specific applications, stand-alone applications, and advertisements. Describing the Java code and its extension for media and digital TV receivers is not possible in this book. However, the reader can refer to a number of sources for further knowledge. The Java Media Framework is specified in the *Java Media Framework 1.0 Specification* at http://java.sun.com/products/java-media/jmf/forDevelopers/playerapi/packages.html. This specification documents the APIs. A developer's guide describing the Java Media Framework is available in *Java Media Players* at http://java.sun.com/products/java-media/jmf/forDevelopers/playerguide/index.html.

Table 4.3
Use of HTML for Display of Message

```
<HTML>
<BODY BGCOLOR= transparent >
<center>
<FONT SIZE=6 color=EFEFEF>
Tune at 8.30 PM for   Yes Minister   serial.
</FONT>
</CENTER>
</BODY>
</HTML>
```

The DAVIC has defined Java language APIs for digital television. These can be found at ftp://ftp.davic.org/Davic/Pub/Spec1_4/14p09ml.zip in Part 9. The Java TV API definition can be found in Javadoc format at http://java.sun.com/products/javatv/jdocs/index.html.

4.6   Authoring Interactive Content

Authoring ITV content is quite different from authoring a typical Web page. A designer determines the look and feel of ITV, and appropriate page backgrounds are developed. Triggers and links are then added for each specific show. A number of tools are available for encoding EIA-746 triggers.

4.6.1   Tools for Encoding EIA-746 Triggers

A few of the tools for encoding EIA-746 triggers are described below:

TV Navigator for ISP

TV Navigator for ISP supports the EIA-746 standard. EIA-746 enables the insertion of HTML content into line 21 of the VBI. This functionality has been branded as Showsync or TV Crossover Links. The idea is that the HTML content represents a URL related to the concurrent television programming. If Showsync content is inserted into the VBI of a television program or commercial, TV Navigator for ISP's status bar pops up, displaying the Showsync URL. The viewer has the choice of dismissing the status bar or fetching the HTML content associated to the displayed URL (usually a Web site).

Norpak

Encoding EIA-746 triggers into a video stream requires a hardware inserter. For internal development purposes, Norpak has used the TES3 inserter with NABTS software. NABTS software is built into the Norpak inserter to allow it to interface with Mixed Signals TV Link Creator™ 2.0, or any generic terminal program [9, 10].

Mixed Signals

Software tools from Mixed Signals can be used to automate the insertion process. These tools are called TV Link Creator 2.0 and allow presetting of all the HTML insertions before editing a tape. It is possible to preset the triggers to appear at various times during the show. The source tape can be played and synchronized using TV Link Creator 2.0 [11].

4.6.2 Authoring Example

Authoring interactive content includes a two-step process that takes previously encoded captions and adds interactive content through an encoding system and software. The first step is to develop video footage and HTML content. The HTML content and video content are developed simultaneously to ensure close compatibility and seamless operation. The images and text of the Web page can overlay the video, or the video can be embedded into the Web page like an image. This is completed before the actual insertion.

The videotape and HTML content are delivered to an ITV work station running software such as TV Link Creator by Mixed Signals (Figure 4.5). The workstation has inputs for video and time code from the video tape recorder (VTR). The workstation takes in a disk file containing closed captions that were prepared earlier. As the ITV workstation operator views the program, he inserts triggers and links at the appropriate times. (The software resolves conflicts between closed captions and ITV content.) At the end of the process, the operator can either write out a new disk file containing both closed captions and ITV content or encode the combined closed captions and ITV cues into a new tape using a Mixed Signals DV2000 encoder. In any case, the closed caption and ITV data are usually encoded into the vertical interval for later playback. In some cases, the disk file is played out to air through a caption server during playback of the show [12].

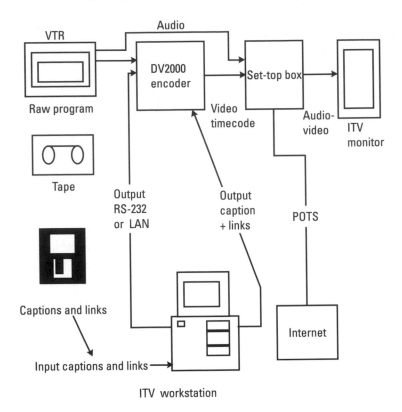

Figure 4.5 Authoring interactive content.

There are other ITV systems available, such as the one produced by Wink Communications. Wink Communications has been working on ITV solutions for many years. Its current focus is on interactivity and e-commerce solutions for television. Liberate Technologies is another vendor of interactive solutions. New companies are entering the market all the time.

4.7 WebTV

WebTV is one of the most successful and popular ITV systems that integrates the Web with TV. A WebTV system allows users to connect to the Internet easily and inexpensively through a small set-top box, using their television as a monitor. All WebTV systems allow surfing of the Web. Few WebTV systems are capable of delivering enhanced television [13] or satellite compatibility [14]. One model of the WebTV has been designed solely for the Internet [15].

A WebTV system uses a WebTV browser, which facilitates Web content delivery on the TV monitor in esthetic form. Users can browse Web pages and read e-mail. The WebTV onscreen interface has been designed specifically for television and is simpler and easier to use than computer-based Web browsers. Users navigate the Web with a selection box controlled by a handheld remote control or keyboard. However, WebTV has a few limitations. Only one window can open at a time. Users cannot download files that require an external application to be run, such as Adobe Acrobat or Microsoft Word files. However, some file formats (such as .zip) have been integrated into the browser and will work.

There are three main models of WebTV boxes, namely Classic [16], Plus [17], and DishPlayer [18]. The Classic model is the oldest of the three and offers only the ability to surf the Web over a TV. The Plus model is the second-generation box, which provides a slightly more advanced browser and introduces interactive TV functionality. The DishPlayer is essentially a Plus box built into a Dish Networks satellite receiver, with the added benefit of digital video recording (DVR) capabilities [19]. A comparison of DishPlayer with TIVO (a personal TV service by TiVo) is available on the Web [20].

4.7.1 WebTV Browser Navigation

WebTV users do not use a cursor and mouse to browse the Web. This is because WebTV users navigate Web sites from a couch or recliner rather than a desk. A mouse, which needs desk space to function, cannot be used as a navigation tool. As such, WebTV users navigate the Internet with a yellow selection box controlled by arrow buttons on their keyboard or remote control. This selection box allows users to move through the hyperlinks on a Web page, as demonstrated in the two screen shots of WebTV (Figure 4.6 and Figure 4.7). This type of selection box offers both precision and a fast learning curve.

4.7.2 Web Page Design

As discussed earlier for ITV design, a number of features are to be incorporated while programming for WebTV. Although WebTV uses a selection box for navigation, it may not always go where expected on complex pages. In addition, some types of image maps can be difficult to navigate. The first

Figure 4.6 Screen shot showing the selection box surrounding "Getting there."

Figure 4.7 Screen shot showing the selection box on "Step 1: Where to start."

principle is to keep all links grouped in the same location. This facilitates easy navigation.

4.7.2.1   Content Design for WebTV

Supported Code and File Formats

The most important part of designing any page is ensuring that users will actually be able to read and use the content on the site. HTML is simply a set of commands that are put in the body of the text that tells WebTV what to do. WebTV-specific HTML tags and attributes are available on site [21]. Appendix 4.C provides the details of the syntax, attributes, also various supports, namely HTML, DHTML, JavaScript, image, media, flash, and other programs, that WebTV provides for authoring content for WebTV [22–26].

Inserting ITV Links

ITV programming for WebTV is designed by introducing relevant Web content into the television experience, not by simply inserting a random TV show into a Web page. A view (view:tv) trigger attribute in the ITV link is used to specify the content Web page. This tells receivers to display this particular content as video. A tv: URL in the HTML of the Web page is used to specify how the video component appears on the page. When received by WebTV for Windows, all of the attributes except the URL and checksum are ignored, but the target Web page is still displayed in a new browser window.

ITV links work on both televisions and computers. To view ITV links on the television, viewers need to use the WebTV Internet receiver (a set-top box that uses the WebTV® Plus service). To see them on a computer, viewers need to use Microsoft® WebTV for Windows® (a feature of Microsoft Windows 98) with a compatible TV tuner card. ITV links basically work the same on both receivers; however, there are a few differences in how they are handled.

Trigger Matching

Trigger matching enables the WebTV client to send a script contained in an ITV link (called a trigger) to a set of Web pages, instead of to a single Web page. In order for the client to execute the script on successive pages, it must determine that a match exists. The client determines that a match exists if (1) the final character in the trigger is an asterisk (*), and (2) the URL of the top-level page of the current ITV program matches up to the position of the asterisk.

4.7.2.2 Design Aspects

Content Display

All WebTV systems in North America and Japan, which use the NTSC television standard, display Web pages in a fixed 544 by 372 screen space. Pages may scroll vertically but not horizontally. Pages that are wider than 544 pixels are scaled to fit that width. While users have no trouble scrolling through a long page, it is good idea to make short pages fit the 372-pixel boundary. Users will not like to scroll to find only an extra sentence or two.

Text on Page

The default font for WebTV users is Helvetica, approximately 18 points in size. The only other font available is Monaco, which is used as a fixed-width font for tags that require one, such as `<code>`. Due to the smaller screen resolution and larger font size, less text fits on a television screen than a computer monitor. This makes it necessary to be very concise when creating content. Large paragraphs should be broken into groupings of two or three sentences. Small text sizes in HTML or embedded in graphics are to be avoided.

Text in Graphics

The display on a traditional television screen is not as crisp as on a computer monitor. Graphical text should be more than 16 points in size to ensure legibility on a television screen. Bold text and high contrast between text and background color also enhances legibility (Section 4.4).

Pages and Files

Pages and files should be as small as possible. It is difficult to recommend an upper boundary on page size. However, a limit of 250K for Web pages may be safe. Of course, the main problem with a 250K page is not the browser's technical ability to display it; it is the amount of patience the user has for lengthy downloads. Like everything else on the Web, small size is a virtue. MPEG and audio files should also be smaller than 400K. This will result in quicker downloads on the viewer's receiver.

Page Titles

The title of the page is provided in the `<title>` tag. Twelve characters of a page's title are visible in the recent panel. Twenty characters of a page's title are visible in the contacting publisher panel (shown when the browser is going to a new page). Twenty-five characters of a page's title are visible from the favorites screen. Thirty-five characters of a page's title are visible in the WebTV browser's status bar. More than 100 characters of a page's title are visible in the info panel of the WebTV browser. These are maximum characters that could possibly be used. It is better to create titles so that the most relevant and distinguishing information is in the first 12 to 20 characters. Titles longer than 35 characters are to be avoided.

Colors

A dark background with light text should be used for best readability on television.

Tables or Frames

The WebTV browser converts frames into tables. The large font of a WebTV browser allows only a few words per line. While slicing up large images can dramatically improve the download time of pages, large spliced images can lose integrity, or break, when viewed through a WebTV browser. It takes only a few small additions to the code to rehabilitate these broken images without visible differences in other browsers [27]. The WebTV browser, on the other hand, seamlessly integrates image maps into the page navigation so image hot spots appear as regular links. Client-side image maps (hot spots are defined in the .html file) with rectangular hot spots are the best type of image maps to use for a WebTV browser. The hot spots appear as regular links, clickable with the user's selection box.

WebTV users employ a selection box to surf the Web. While this selection box works seamlessly on client-side image maps with rectangular hot spots, any other type of image map can be inconvenient to use.

Forms

Forms are to be kept simple. Long pull-down menus (tedious to use) and long values on buttons (they may get truncated) should be avoided. The WebTV browser can be confused by extraneous or bizarre code. Complicated or inconsistent forms can also confuse WebTV users.

4.7.3 WebTV Proxy Server

Proxy servers allow WebTV to speed up users' Web sessions. By caching Web content, proxies catch the Web content and deliver popular pages without needing to retrieve a unique copy from the Internet each time. Proxy servers also tweak individual pages to ensure that they display properly on a television screen.

To speed surfing, a WebTV terminal issues requests through a number of proxies; anywhere from two to a dozen. This results in a correspondingly large number of IP addresses for a single WebTV user during a single session for nonsecure connections. If a connection is made through a secure socket layer (SSL), the WebTV users bypass the proxy and have a static IP until an HTTP connection is resumed.

4.7.4   SSL: Securing Transactions

Secure access is essential for any kind of electronic commerce and for exchanging personal information. This requires a secure Web server running SSL and a client that is SSL compliant. The WebTV client supports 128-bit SSL encryption. WebTV supports Versions 2 and 3 of SSL. Details are provided in Appendix 4.C.

All transactions are point-to-point for security and privacy reasons. When a viewer connects to a secure site, the connection does not go through the WebTV service. The viewer is directly connected to the concerned server. Figure 4.8 illustrates the difference between secure and nonsecure connections on WebTV. It can be seen that in a secure connection to Grindlay Bank, the viewer's set-top box is directly connected to the bank. In the nonsecure connection, the consumer passes through the WebTV service to get connected to the bank. In this case, the proxy server can cache the transaction details.

4.7.5   WebTV Emulator

WebTV Viewer [28] is a free program that emulates the WebTV environment on a PC or Macintosh. The program enables browsing the Web in

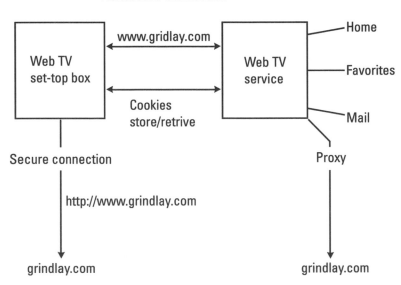

Figure 4.8 Secure and nonsecure connections.

much the same way as a WebTV user. The WebTV Viewer reproduces layout changes caused by the way WebTV displays pages, such as narrowing of graphics and tables to the WebTV screen width of 544 pixels. The WebTV Viewer does not reproduce the blurring effect of a TV screen or its treatment of color. WebTV Viewer can be used to check a Web site for layout problems and to preview Web pages under development. The viewer can be downloaded from the Web site [29].

The WebTV Viewer is a useful tool for WebTV developers. It provides a general idea of how the sites will display on WebTV. It is the easiest method for testing look-and-feel issues. But it also has many limitations. Many of the readability issues (such as text size and color contrast) are caused specifically by the content being displayed on an NTSC television screen, so these sorts of problems cannot be detected using the emulator on the PC monitor. For example, straight black text (#000000) on a straight-white background (#FFFFFF) will read just fine on the emulator but will be difficult to read on a TV screen, since the high-contrast colors appear to bleed together at the edges. In addition, the emulator only simulates the display functionality of the WebTV browser and does not accurately represent the back-end workings of the browser and operating system, as they apply to the functionality of a site. This usually comes into effect when dealing with SSL, cookies, or higher-level JavaScript.

Final Test

The final test of the pages should be made on a real WebTV system. Many retail outlets [30] that have WebTV systems on display provide "hands-on" time.

4.7.6 ITV Programming Examples

Examples of ITV shows complete with code and graphics are available on the Web. "Only Hits" is a simple ITV show designed specifically for a WebTV Internet receiver. The site is designed around the business model that describes a viewer most likely to make an impulse purchase of a music CD after watching the video on a music video channel and hearing a hit from that CD. It includes all the program's components, including the ITV link, Web content, and CDF file [31]. Another example simulates a presidential election debate in which viewers provide their opinions on a specific topic and then the results are tabulated [32]. Anyone can use these examples to gain an understanding of how to build ITV shows.

Another site maintained by a WebTV Networks engineer has other examples to run on a WebTV Internet receiver (it is not intended to work on a computer) [33]. These examples include button overlay, banner overlay, flash, scrolling frames, marquee, embedded animation, and other programs [34]. The site also provides ITV JavaScript test cases [35]. ITV templates are also available for developers [36]. These include news and sports templates, as well as others. The sports interactive experience is composed of several transparent overlays and pop-up controls. The application uses a pop-up menu that provides access to game updates and player statistics on demand. An example of creating selectable pop-ups that appear over full-screen video or other pages is also available [37]. The template provides the basic functionality needed to create selectable pop-ups that appear via triggers.

4.8   ITV Development Resources

A Web site maintains the details of the hardware that can be used to encode ITV links, software, and other tools for producing ITV shows, as well as the names of production companies that can help with the process [38–40].

Links are used to trigger interactive programming. The links have a specific syntax, and some of the rules on abbreviations can be difficult. Moreover, after the programming is over, a checksum is required for the whole string. A tool is available that helps generate ITV links [41].

4.9   Summary

Content plays the most important role in making ITV popular. ITV is being used to provide personalized services and give users control over what they watch. TV has the capability of keeping millions of eyeballs glued to a program. Many technological and behavioral changes are transforming television, such as the development of megachannel cable TV systems, menu-driven user interfaces for TV (in which viewers can navigate through large-capacity cable TV systems), and increased channel changing by viewers. The design and placement of commercials has been affected by these trends. In an effort to reduce channel changing, many stations and networks no longer place commercials at the end of local news shows but now prefer "hot switching" or "seamless programming," where they cut directly to the opening of network programs. The potential of t-commerce can be harnessed by imaginative and innovative programming. Therefore, content creation for ITV is not merely integrating Web pages into TV or vice-versa. Several issues aside from coding, including design and aesthetic, need to be addressed regarding how content is created. In this chapter, we examined several aspects of content creation for ITV. We also saw how the URI scheme allows referencing of various resources. The use of triggers and their syntax as per AVTEF standards were described. WebTV as a special example of ITV was also examined.

References

[1] Hunter, J. R., H. Lau, and D. J. White, "Enhanced Television Service Development," BBC Research & Development, United Kingdom.

[2] http://www.cardtech.faulknergray.com/julart.htm, March 2001.

[3] [itvt] Issue 3.34, April 4, 2001.

[4] http://www.atvef.com/library/spec1_1a.htm#746A, April 2001.

[5] http://global.ihs.com/, April 2001.

[6] http://www.tiaonline.org/standards/search_n_order.html, April 2001.

[7] http://www.w3.org/TR/NOTE-CDFsubmit.html, April 2001.

[8] http://www.atvef.com/, April 2001.

[9] http://www.norpak.ca/tes3wst.htm, April 2001.

[10] http://www.norpak.ca/Nabts.htm, March 2001.

[11] http://www.mixedsignals.com/, April 2001.

[12] Gilmer, B., "Interactive Television for Terrestrial Broadcasters," at http://www.broadcast-engineering.com/html/2000/september/digitalHandbook/computer&Networks.htm, April 2001.

[13] http://www.webtv.net/products/plus/, May 2001.

[14] http://www.webtv.net/products/satellite/, May 2001.

[15] http://www.webtv.net/products/classic/, May 2001.

[16] http://www.webtv.net/products/classic/specs.html, May 2001.

[17] http://www.webtv.net/products/plus/specs.html, May 2001.

[18] http://www.webtv.net/products/personaltv/specs.html, May 2001.

[19] http://www.webtv.net/products/personaltv/ptv.html, May 2001.

[20] http://www.evolt.org/article/Testing_Your_Sites_for_WebTV_Browsers/20/5409/, April 2001.

[21] http://developer.webtv.net/design/authoring/authoring/html/WebTV_extentions.html, May 2001.

[22] http://developer.webtv.net/authoring/faq/Default.htm, April 2001.

[23] http://developer.webtv.net/design/authoring/Default.htm, April 2001.

[24] http://developer.webtv.net/design/authoring/html/Default.htm, April 2001.

[25] http://developer.webtv.net/design/authoring/javascript/javascript.htm, April 2001.

[26] http://developer.webtv.net/design/authoring/Default.htm, April 2001.

[27] http://developer.webtv.net/authoring/splice/, April 2001.

[28] http://developer.webtv.net/design/tools/viewer/instructions/default.htm, May 2001.

[29] http://developer.webtv.net/design/tools/viewer/default.htm, May 2001.

[30] http://www.webtv.net/products/lookup/find_form.asp, May 2001.

[31] http://developer.webtv.net/itv/examples/music/default.htm, April 2001.

[32] http://developer.webtv.net/itv/examples/debate/default.htm, April 2001.

[33] http://itv.webtv.net, April 2001.

[34] http://itv.webtv.net/examples/index.html, April 2001.

[35] http://itv.webtv.net/matt/testcases/index.html, March 2001.

[36] http://itv.webtv.net/templates/index.html, April 2001.

[37] http://itv.webtv.net/templates/popup2/popup2_index.html, April 2001.

[38] http://developer.webtv.net/itv/whatis/main.htm, April 2001.

[39] http://developer.webtv.net/itv/resources/default.htm, April 2001.

[40] http://developer.webtv.net/design/checklist/main.htm, April 2001.

[41] http://developer.webtv.net/itv/tvlink/default.htm, February 2001.

Selected Bibliography

Arnold, K., and James Gosling, *The Java Programming Language, Second Edition* (ISBN: 0-201-31006-6), http://java.sun.com/docs/books/javaprog/secondedition, April 2001.

Bradner, S., RFC 2119—Key Words for Use in RFCs to Indicate Requirement Levels, http://info.internet.isi.edu/in-notes/rfc/files/rfc2119.txt, March 1997.

Berners-Lee, T., R. Fielding, and L. Masinter, RFC 2396—Uniform Resource Identifiers (URI): Generic Syntax, http://info.internet.isi.edu/in-notes/rfc/files/rfc2119.txt, August 1998.

Daniel, R., and M. Mealling, RFC 2168—Resolution of Uniform Resource Identifiers Using the Domain Name System, http://info.internet.isi.edu/in-notes/rfc/files/rfc2168.txt, June 1997.

Finseth, C., "Applications List," http://list-w3.org/ArchivesPublic/www-tv/1998OctDec/0224.html, December 1998.

Gosling, J., B. Joy, and G. Steele, The Java Language Specification, (ISBN 0-201-63451-1) at http://java.sun.com/docs/books/jls, April 2001.

Hoschka, Philipp, W3C TVWeb Interest Group, http://www.w3.org/TV/TVWeb, March 2001.

http://itv.webtv.net/templates/, April 2001.

http://net4tv.com/, April 2001.

http://partners.liberate.com/contentpartners/kitchen/cdk/index.html, February 2001.

http://ruel.net/tv/index.htm, February 2001.

http://www.dev4tv.com/docs.htm, April 2001.

http://www.ed.ac.uk/~ejuv19/dnhom4.html, December 2000.

http://www.strategic-e.com/interactiveTV.htm, December 2000.

http://www.teleport.com/~samc/cable1.html, October 2000.

http://www.w3.org/TR, April 2001.

http://www.w3.org/TR/1999/NOTE-TVWeb-URI-Requirements-19991019, December 2000.

http://www.w3.org/TR/1999/NOTE-TVWeb-URI-Requirements-19991021, December 2000.

http://www.w3.org/TR/TVWeb-URI-Requirements, December 2000.

Interactive Channel Case Study, Making Interactive Television Work with the Web, http://www.opentv.com/industtry/tvexpert/s_casestudiesIC.html, October 2000.

Java Platform 1.1 Core API Specification, http://java.sun.com/products/jdk/1.1/docs/api/packages.html, March 2001.

Lindholm, T., and F. Yellin, The Java Virtual Machine Specification, (ISBN 0-201-63452-X) at http://java.sun.com/docs/books/vmspec, January 2001.

Masinter, L., et. al, Guidelines for New URL Schemes, http://info.internet.isi.edu/in-drafts/files/draft-ietf-urlreg-guide05.txt, March 1999.

Moats, R., RFC 2141—URN Syntax, http://info.internet.isi.edu/in-notes/rfc/files/rfc2141.txt, May 1997.

Patrick, C., and R. Lee, The Java Class Libraries, Second Edition, Volume 2, (ISBN 0-201-31003-1), http://java.sun.com/docs/books/chanlee/second_edition/, January 2001.

PersonalJava API Specification, Version 1.1, http://java.sun.com/products/personaljava/spec-1-1/pJavaSpec.html, April 2001.

SSL Java Standard Extension to JDK 1.1, http://java.sun.com/security/ssl/API_users_guide.html, October 2000.

Threads Starting with Messages 0040, 0041, and 0046, http://lists.w3.org/Archives/Public/www-tv/, October/November 1998.

TV-Web Mail Archives, http://list.w3.org/Archives/Public/www-tv/, February 2001.

Appendix 4.A

Table 4.A.1
ITV Link Attributes and Their Syntax

Syntax	
URL	The URL is enclosed in angle brackets (for example `<http://www.airkode.net/ main.html>`). Although any URL can be sent in this syntax, ATVEF content level 1 only requires support for `http:` and `lid:` URL schemes.
	The syntax of the `lid:` URL is as follows: `lid://{namespace-id}[/{resource-path}]`
Attribute-Value	
`[name:string]`	The name attribute provides a readable text description (for example, `[name:DDInternational]`). The string is any string of characters between 0x20 and 0x7e except square brackets (0x5b and 0x5d) and angle brackets (0x3c and 0x3e). The name attribute can be abbreviated as the single letter "`n`" (for example, `[n:DDInternational]`).
`[expires:dateT-time]`	The expire attribute provides an expiration date, after which the link is no longer valid (for example, `[expires:2001122-3T000000]`). The syntax is `yyyymmddThhmmss`. The time conforms to the ISO-8601 standard. Letter "`T`" separates the date from the time. When no time is specified, expiration is at the beginning of the specified day. The expires attribute can be abbreviated as the single letter "`e`" (for example, `[e20011223]`).
`[script:string]`	The script attribute provides a script fragment to execute within the context of the page containing the trigger receiver object (for example, `[script:showsports()]`). The string is an ECMAScript fragment. The script attribute can be abbreviated as the single letter "`s`" (for example, `[s:showsports()]`).

Table 4.A.1 (continued)

Attribute-Value	
[checksum]	The optional checksum must come at the end of the trigger. EIA-746A requires the inclusion of a checksum to ensure data integrity over line 21 bindings. In other bindings, such as IP, this may not be necessary and is not required.
	Checksum is a method of verifying if all of the information (data bits) contained in a link is retrieved by a receiver. checksum is a total count of the number of data bits contained in a transmission. This number is included as part of the transmission so the receiver can verify that the correct number of bits arrived. When the numbers match, it is assumed that the transmission was successful.
[script:string]	
	Important: After calculating checksum, the following cannot be done:
	• Change the case of any text;
	• Add or remove spaces;
	• Format the string in any way.
	If there are changes in any of the link attributes, checksum will need to be recalculated.
	Note: Other details are available at http://www.atvef.com/library/spec1_1a

Appendix 4.B

Enhanced Television Developers, Producers, and Consultants

Alliance Atlantis Communications, Inc. (www.allianceatlantis.com)

Alliance Atlantis Communications, Inc. is a creator, producer, distributor and broadcaster of filmed entertainment with significant ownership interests in seven Canadian specialty television networks. The company's principal business activities are conducted through three operating groups: Television, Motion Pictures, and Broadcasting. Headquartered in Toronto, Alliance Atlantis operates offices in Los Angeles, in the United States; Montreal and Vancouver, in Canada; London, in the United Kingdom; Sydney, in Australia; and Shannon, in Ireland.

Back Alley Film Productions

Back Alley Film Productions is mainly involved in television and documentary filmmaking. Comprised of Janis Lundman and Adrienne Mitchell, the executive producers and creators of *Drop the Beat*, the company's most recent work was the critically acclaimed *Straight Up*, about an eclectic group of Toronto teenagers. Back Alley Films counts among its credits *Law and Order* and the award-winning feature documentary *Talk 16*, which aired on television around the world and was released theatrically in the United States, Australia, and Canada.

COM One

COM One, a European manufacturer of communications products, uses OpenTV Device Mosaic in its SurfTV, a TV Internet decoder. France Telecom recently announced it will use SurfTV to enhance the Minitel information system.

DIRECTV, Inc. (www.directv.com)

DIRECTV is a leading digital satellite television service provider with more than 9 million customers. DIRECTV, Inc. is a unit of Hughes Electronics Corp. Hughes Electronics is the world's leading provider of digital television entertainment, satellite, and wireless systems and services.

DOMITEL (http://www.cet.hut.fi/domitel)

DOMITEL develops a supporting telematic learning environment for home learners using interactive CATV networks. It offers a range of interactivity from one-way video delivery combined with interactive teletext to fully two-way CATV allowing for multimedia applications and videophony.

ExtendMedia (www.extend.com)

ExtendMedia is a production company focused on creating interactive content for the Internet, television, and wireless convergence market. By collaborating with clients, ExtendMedia manages the integration and delivery of interactive content, extends their clients' brands to a wider audience, and creates new business models and e-commerce revenue streams. Its market-leading, high-profile partners and investors include Creative Artists Agency, BCE, and Alliance Atlantis, to name a few. Entertainment, broadcast, cable, and telecommunications clients like PBS, New Line Cinema, Eyemark Entertainment, Discovery Health, and Cablevision depend on

ExtendMedia for their emerging interactive needs. Founded in 1991, ExtendMedia is headquartered in Toronto, with offices in New York and Los Angeles.

Grudig, Loewe, and Schneider Rundfunkwerke

Grudig, Loewe, and Schneider Rundfunkwerke, three of Germany's consumer electronic manufacturers, are currently working with QNX Software to develop Web-enabled television sets. All three companies are using OpenTV Device Mosaic as the browser solution for their products.

IDI Media Interactive Production Services (http://www.iditv.com)

IDI Media provides a turnkey solution for production of the interactive and on-line elements of television programming. IDI also develops enhancements and applications programs for the WebTV platform.

Interactive Channel

Interactive Channel uses OpenTV Remote Mosaic technology, OpenTV Prism, and SurfWatch technologies to provide high-quality Internet access and interactive services to consumers through their cable TV.

Internet Television Network (http://intv.net)

The Internet Television Network provides streaming media programming on the Web, producing and delivering business-to-business and business-to-consumer content worldwide.

Liberate (http://www.liberate.com)

Liberate develops the software platform for interactive TV. Cable, satellite, and telecommunications network operators use Liberate's software to roll out interactive digital services on a variety of devices from low- to high-end TV set-top boxes.

Mixed Signals Technologies, Inc. (http://www.mixedsignals.com)

Mixed Signals is the industry's leading provider of ITV technology, including equipment, expertise, and service.

NetChannel (http://www.netchannel.net)

NetChannel is an Internet and information service designed to extend the television experience by delivering personally relevant information and entertainment through the TV. The company uses OpenTV's SurfWatch technologies to ensure that families are able to safely explore the Internet without encountering objectionable material.

Netmèdia Advanced TV Resources (http://www.iua.upf.es/~ctomas/ctp30.htm)

Netmedia provides advanced TV resources and also conducts research and development.

OpenTV (http://www.opentv.com)

OpenTV provides software solutions to develop and deliver interactive services via digital satellite, cable, and terrestrial broadcast. OpenTV's SurfWatch technologies ensure that families are able to safely explore the Internet without encountering objectionable material. OpenTV is now installed in more than 7.8 million digital set-top boxes worldwide. Twenty digital TV network operators have deployed OpenTV systems, including 6 cable, 12 satellite, 1 terrestrial and 1 multichannel multipoint distribution service (MMDS) operator. Two digital satellite TV network operators, British Sky Broadcasting in the United Kingdom and Television Par Satellite in France, have deployed more than one million OpenTV systems on a cumulative basis. Thirteen digital set-top box manufacturers have shipped OpenTV set-top boxes, with 10 manufacturers shipping more than 300,000 units each.

A leading provider of digital set-top boxes for cable and satellite, Nokia turned to OpenTV to incorporate Web technology into its next-generation digital cable, satellite, and terrestrial set-top boxes. The OpenTV Device Mosaic Web browser was modified and used by Nokia as an HTML engine for developing interactive and on-line TV services. The inclusion of an HTML engine makes it possible to create a host of advanced ITV services, including: interactive advertising; home shopping; electronic programming guides; VOD, and pay-per-view ordering; all types of electronic commerce, including banking; and traditional Web browsing, Internet access, e-mail, and chat services.

PowerTV (http://www.powertv.com)

PowerTV, an ITV application developer financed by Scientific-Atlanta, uses OpenTV Device Mosaic to allow its TV operating system to work with the Web. Currently, PowerTV is developing software for a number of U.S. operators implementing advanced digital HFC TV systems, including TimeWarner, Pacific Bell, and BellSouth, as well as the international IMMXpress consortium composed of Siemens, Sun Microsystems, and Scientific-Atlanta.

Screamingly Different Entertainment, Inc. (http://www.screaminglydifferent.com)

Screamingly Different Entertainment, Inc. provides turnkey solutions to develop, produce, and deliver interactive, enhanced, and convergence content to the entertainment industry.

Scientific Atlanta (http://www.sciatl.com/nav/html/top/loframe.html)

Scientific-Atlanta provides interactive telecommunications products and services. Broadband Access products help cable operators deliver more channels and exciting services, such as the Internet, VOD, e-commerce, and more. Content Distribution (PowerVu Product Family) products help programmers, broadcasters, and service providers deliver more video channels and IP applications over their satellite transponders to cable operators, businesses, and consumers.

StarSight Interactive TV Program Guide Service (http://www.starsight.com)

StarSight is an interactive on-screen TV program guide service with one-button VCR programming. StarSight is currently available in many TVs, VCRs, TVCRs (units combining a TV and VCR), cable set-top converters, satellite receivers, and a stand-alone receiver that works with broadcast or cable signals. StarSight features are also available in many DSS systems.

The Kiss Principle Inc. (http://www.thekissprinciple.com)

The Kiss Principle, Inc., is the developer of conversational interfaces for TV and other devices that enhance programming, sponsorship, and advertising. Conversational interfaces turn chat into a fun interface to information and services by using prescripted dialog produced in conjunction with TV producers. The Kiss Principle's patent-pending technologies provide supplemental entertainment, community, and support to TV programming.

Thomson Consumer Electronics (www.powertv.com/press/thompson.html)

Thomson Consumer Electronics offers the RCA Network Computer, an affordable Internet access product that lets consumers use their TVs to surf the Web. OpenTV Professional Services developed an EPG that is provided as part of the personalized Internet service included in the RCA NC. Through the EPG, viewers have a single interface that seamlessly combines broadcast programming with related Web content.

THOMSON multimedia (http://www.thomson-multimedia.com)

THOMSON multimedia (Paris Sicovam: 18453, NYSE: TMS) is one of the world's largest producers of consumer electronics. The group has five principal activities: Displays and Components, Consumer Products, New Media Services, Digital Media Solutions, and Patents and Licensing. THOMSON multimedia deals with consumer products such as TVs, VCRs, audio systems, digital decoders, DVD players, and professional video equipment under the popular RCA and THOMSON brand names.

VEIL Interactive Technologies (http://www.veilinteractive.com)

VEIL™ Interactive Technologies uses a technology called VEIL, which stands for video encoded invisible light. VEIL allows inserting extra, digital information into an ordinary TV or video picture. This extra information works as a trigger that causes special interactive opportunities to happen. There are many different ways to use the VEIL encoded information. The FCC reviewed and approved VEIL in 1987. The U.S. Patent Office issued a patent for VEIL on February 21, 1989. VEIL is compatible with all current digital compression and videotape standards.

Visiware: ITV developer (http://www.visiware.com/site/en/accueil_flash.htm)

Visiware offers on-line video games and ITV under OpenTV digital set-top boxes. When the author checked the site, *Arabian Nights, Episode 7* was available.

WebTV (http:// www.webtv.com)

WebTV, a Microsoft subsidiary that is dedicated to bringing Internet access to TV consumers, uses OpenTV SurfWatch technology to ensure that its Internet service delivers content appropriate for all audiences.

Welcome To The Future, Inc. (http://www.wttf.com)

The company provides Internet and wireless ITV.

Wink Communications, Inc. (www.wink.com)

Wink Communications, Inc., is the U.S. leader in enhanced broadcasting and one-click commerce over television. The company is based in Alameda, Calif., and provides an end-to-end system to enable the delivery of ITV entertainment and the collection of transactions generated by TV viewers. The TV feature, called Wink Enhanced Broadcasting, allows viewers an easy and convenient way to interact with programs and advertisements using their remote control while they continue to watch TV. Wink is free to use and requires no upgrades or additional hardware in the home for viewers that have a Wink-enabled digital satellite or cable set-top box.

Wink's Enhanced TV was introduced in the United States in June 1998 and is currently available to more than 300,000 households in a variety of locations throughout the country. Wink has been available in Japan since October 1996. A complete list of industry leaders who have adopted the Wink technology and additional information regarding Wink Communications is available at www.wink.com.

WorldGate (http://www.wgate.com)

WorldGate, a leading provider of Internet cable service, incorporates OpenTV browser technology into its WorldGate TV on-line service. The service is provided in collaboration with four of the country's largest cable operators—Cablevision, Adelphia, Comcast, and Charter Communications. The service allows customers to access the Internet through their TVs without the use of a PC, a modem, or additional in-home equipment.

Appendix 4.C

4.C.1 WebTV Link Attributes and Their Syntax

WebTV links can have any combination of a number of attributes. The attributes are url, type, name, script, expires, view, checksum. The general format for triggers is a required URL followed by zero or more attribute-value pairs and an optional checksum:

```
<url>[attr1: val1][attr2:val2]...[attrn:valn][checksum]
```

URL

Description
The Web site address of the ITV content.
    Required.

Example

```
<http://www.webtv.net/sports/itvcontent/>
```

TYPE

Description

Used to categorize ITV links, this attribute also shows the TV viewer any links that appeared during a program or time period. A type attribute is added by using either the word type or the abbreviation t.

Optional (CEMA permits omission of the word type).

Attributes

Program—the current program;

Network—the broadcast network;

Station—the local station;

Sponsor—a commercial sponsor of the current program;

Operator—the service operator (for example, cable or satellite).

Examples

[t:o]

[type:operator]

NAME

Description

The name attribute is for the display of the message that TV viewers see on the drop-down panel after they select the ITV link icon. This is limited to 20 or 30 characters, depending on the characters used (a proportional font is used). The name is the only information TV viewers see about the interactivity before playing it.

Optional.

Examples

[n:WebTV Networks]

[name:WebTV Networks]

SCRIPT

Description

The script attribute is the trigger actions on the Web pages that the ITV link references. The value specifies a script fragment that will be sent to the target

Web page and executed. The context for the script is the root document corresponding to the URL specified in the link string.

Optional.

Examples

```
[s:shownews]

[script:shownews]
```

EXPIRES

Description

Specifies the last date that the ITV link is valid. The format of the time conforms to ISO-8601 and is relative to Greenwich Mean Time (GMT). A valid compact format is yyyymmddThhmmss, where T separates the date from the time.

Optional.

Examples

```
[e:20010324T125656]

[e:20010324]

[expires:20010324T125656]
```

VIEW

Description

Specifies how the receiver should display the linked resource to the TV audience.

There are two ways to view Web pages, as TV-related content ([view:tv]) or as a conventional Web page ([view:web]). If not specified, the receiver will default to [view:web].

Optional.

Examples

```
[v:t]

[view:tv]
```

CHECKSUM

Description

Verifies the accuracy of the string and detects data corruption. The `check-sum` is established in the CEMA standard and is required for line 21 triggers only. It is not required for triggers sent in IP packets.

Required.

4.C.2 WebTV Supports

HTML

A reference to understand which HTML elements WebTV supports is available at http://developer.webtv.net/authoring/html/Default.htm. The pdf version of the HTML reference guide can be downloaded from the site http://developer.webtv.net/authoring/html/htmlpdf/html_ref.pdf.

Dynamic HTML in the WebTV Plus Browser

Dynamic HTML (DHTML) is a term used to describe the combination of HTML, style sheets, and scripts that allows Web content developers to animate documents. Cascading style sheets (CSS) are one of three components in DHTML; the other two are HTML and JavaScript (which is being standardized under the name ECMAScript). The three components are glued together with the document object model (DOM), which determines how the JavaScript in an HTML document works in a browser.

The WebTV Plus receiver fully supports DOM level 0 (functionality equivalent to that exposed in Netscape Navigator 3.0 and Microsoft Internet Explorer 3). The details are available at http://developer.webtv.net/authoring/dhtml/. The following additional elements are also supported by WebTV:

- `offsetHeight` property

- `offsetLeft` property

- `offsetTop` property

- `offsetWidth` property

- `disabled` property

- `readOnly` property

- Document.all

- visibility property

Forms

The details about the forms to be used on WebTV and their workings is available at http://developer.webtv.net/authoring/forms/.

HTML Frames for WebTV

WebTV supports frames by breaking them out into a table. This means that all of the content within the frames will be displayed, but no individual frame will remain static. The WebTV browser allocates frame size according to a given frame's content. The details about HTML frames and how these affect the site is available at http://developer.webtv.net/authoring/frames/.

CSS on WebTV

Style sheets are supported only on the WebTV Plus Receiver. The WebTV Classic Internet Terminal does not support style sheets in any form.

Java

The WebTV browser does not yet support Java.

JavaScript

WebTV supports JavaScript based on JavaScript 1.2, as defined by EMCA-262 standard, rev. 2, which can be found at http://www.ecma.ch.

SSL

WebTV supports SSL, with both 40- and 128-bit encryption. RSA, GTE, Thawte—issued by Thawte Consulting, VeriSign—Server certificates in the RSA Secure Server trust hierarchy or Class 3 root are supported.

Macromedia Flash

Macromedia Flash allows easy and cost-efficient creation of animation and sound to Web pages. WebTV® supports the use of Flash as a tool for creating Web sites that work well on both PCs and WebTV systems. The fast downloads for animated content made possible by Flash are a natural fit for the WebTV audience. All WebTV systems now support Flash 3. Details can be found at http://developer.webtv.net/authoring/flash/Default.htm.

Media

Video

WebTV users can watch video clips on their TV screens. WebTV supports MPEG-4.3 (through Windows Media Player), MPEG-1, and VideoFlash. DishPlayer, a WebTV satellite receiver, does not yet support Windows Media Player

Audio

WebTV supports a number of audio formats such as AIFF, 8/16-bit linear; GSM; Macromedia Version 3 flash; MIDI; MOD; MPEG-1, including Layers I, II, and III, as well as the low-sampling frequency (LSF) versions of MPEG-2, including Layers I, II, and III; QuickTime; RealAudio Versions 1.0, 2.0, and 3.0 with bit-rates up to 22 kHz (only 16 kHz on some platforms); Shockwave; RMF; WAVE; and Windows Media Player.

Image

A WebTV system supports 87a and 89a GIF files, including animated and interlaced GIFs, JPEG, TIFF-G3 XBM Macromedia Flash 3, and PNG.

Web Browsers

WebTV developed its own interface, specifically designed for display on TV. Most Web pages successfully tested on Netscape Navigator 4.0+ and Microsoft Internet Explorer 4.0+ will display well on the WebTV system. A browser sniffer script is used to deliver a certain Web page to WebTV viewers and a different page to Internet Explorer or Netscape visitors.

5

Business Models and Potential

5.1 The Marketplace Context

The development of ITV depends directly on several important issues. The foremost issue is the cost to be borne by consumers in acquiring the receiving equipment. The second is the cost of infrastructure to be borne by service providers. We will discuss the cost of building the infrastructure first. Estimates vary enormously from a few hundred dollars per household to a few thousand dollars. Estimates for building the infrastructure at a national level across the globe depends on the existing state of broadcasting, plans for digital transition, the state of cable systems, Internet development, the capability of providing enhanced content, etc. Projections may vary between $100 million and $500 billion. This cost is to be borne jointly by service providers, network operators, and content developers. Other issues such as consumer service management could also affect service charges.

With costs so high and market demand uncertain, clearly there is a significant level of risk for all players that provide ITV services. However, a closer examination reveals that most of the costs associated with building the new infrastructure will have to be incurred regardless of whether any ITV services are offered.

For example, the future being invariably digital, most broadcasters and telephone companies will change to digital modes of transmission. Deploying higher bandwidth pipes to end users is a major strategic activity of the telephone, wireless, satellite, and cable industries. This will add to infrastructure capacity. Transition to digital broadcasting is already being planned. Many nations have set the date for switching off analog transmission. HFC (Section 1.6.3) and fiber-optic cables will have to be laid to meet the demands of broadband requirements. This means that incremental costs of building a network for ITV are a modest addition to normal replacement and upgrading costs of existing networks. For cable operators, incremental costs are greater but they can scale back investment if necessary. Further, although initial capital investment costs are high, subsequent operating and maintenance costs are likely to be lower. The life-cycle cost of ITV services are likely to be only marginally higher in terms of cost per channel and the type of services and business potentials. A number of companies have already planned substantial investment for developing the infrastructure. Details are provided in Appendix 5.A.

The cost issue for consumer equipment is most important and has agreater impact on ITV's success. We discussed earlier that consumers diverted their subscription to movie-on-demand from other services (Section 2.2.1.1). The gain for one type of service was at the expense of other services. In the broadcast arena, consumers prefer to use their equipment for a longer period, typically 8 to 10 years. In the case of PCs, consumers often have to change their equipment after 2 to 3 years due to rapid technological advancement and resultant obsolescence of hardware and software. Another aspect concerns the form of public broadcasting services, where broadcasters provide a number of channels for free. In many countries, TV receiver licenses have also been waived. Even in the United Kingdom, charges for digital multiplexers for public broadcast services are at a differential rate. This is because broadcast is considered a way to provide education and information to the masses.

In broadcasting, efforts are made to use technology in such a manner that the complexity in the receiving equipment is lessened and it is affordable to people at large. This holds true for ITV as well. Cost issues associated with the design of a network for ITV and equipment configurations in the home need to be carefully considered. It is possible to provide many interactive and "immersive" services by providing complex receiving equipment at the consumer end. Readers may be aware of earlier experiments where it was possible to control the environment at the receiving end. For example, some type of fragrance could be provided based on the scene on the screen. This necessitated storing the fragrance with the receiving equipment and controlling the nozzle by some mechanical means. This was tried in cinema halls; however, all these experiments did not succeed due to various reasons, the foremost being the complexity of the end equipment. ITV designers have to decide whether to place greater power and memory in the set-top box that attaches to a TV or to place greater processing power in the network that will transmit services to the home. A powerful set-top box is attractive to many designers who like the TV to have the same processing capabilities as a high-end PC. But this would increase the cost of the set-top box. The pertinent question is who will pay for it. Consumers may pay for the set-top box at the time of initial purchase. The other alternative is that system operators may provide the box but charge monthly rental fees. In both cases, however, it is the consumer who pays for the box, and the question of affordability arises.

In the other model, it is possible to provide the same capabilities to a less powerful and less expensive box if the network has sufficient processing power [1]. This is an important issue, given the market experience of consumers who have been reluctant to spend hundreds of dollars on dedicated terminals that have a single function. It may be recalled that videogame units have a single function but are relatively inexpensive. VCRs and PCs are more expensive but support multiple uses.

Cost Factor

In the early 1990s, it took months or years to develop proprietary applications for ITV tests and trials. The cost of infrastructure and programming were obviously very high. ITV developers, using Internet tools, can do essentially the same things today in a matter of days or weeks at a fraction of the cost. This is because a digital infrastructure is now in place for every delivery platform, such as terrestrial broadcast in the form of DTT, satellite broadcast, and delivery through cable. Digital set-top boxes are becoming available at an affordable price. Third-party ITV development tools are available. There are a lot of ITV developers, content creators, and programmers. Content creators are producing imaginative and innovative ITV programs. Open standards, including HTML, TCP/IP, ATVEF, DOCSIS, and DVB-MHP, are getting in place. The Web sites of content developers or other companies connected with ITV are provided in Appendix 5.B. All these developments have made this the right time and environment to develop ITV at affordable prices for both consumers and service providers.

5.2 ITV Drivers

The key drivers of ITV services are the worldwide transition to digital broadcast, Internet commerce, optic fiber and broadband growth, TV portals, viewers' preferences, and compelling content. Let us first examine the role of these key drivers. In Section 5.3, we will discuss leading world markets and the projected growth of these services.

5.2.1 Digital Broadcast

Broadcasters are the prime gatekeepers of ITV services to the home. The critical factor in the growth of the ITV market will also depend on broadcasters, who will really control the user's access and quality of ITV service. This differs from access to interactive services on the Web, where the user is actively involved in the innovation process as services develop and evolve.

DTV broadcast facilities have numerous inherent advantages in the infrastructure marketplace for serving client devices running on diverse platforms. These advantages include broadcasting's in-place deployment of analog and increasingly DTV transmission facilities, wireless signal propagation with scheduled and ad hoc regional, national and international networking, low incremental costs to enter and operate in the bandwidth marketplace, ubiquitous and wireless distribution in a local market, and the ability to interoperate with a variety of back-channel technologies for full-duplex operations. DTV is capable of providing high-quality multichannel, multimedia service. Digital TV facilitates building an information society. Programs and data can be broadcast to a TV, PC, or a set-top box. Reception equipment is capable of interactivity. It allows downloading of data or programs that can be stored and played at a later time. There are now more than 6 million DTV receivers deployed in the world. Although only a fraction of the 1.4 billion TV households worldwide, the market is growing steadily. Web access via DTV will increase dramatically after 2001. The type of services provided will consist of noninteractive TV, proprietary ITV, and restricted and unrestricted Web access. Datamonitor predicts there will be 67 million DTV subscribers on all platforms by 2003. The OpenTV system on SkyDigital is generating $1.6 million in revenues every week and is expected to reach $20 billion annually by 2004. Interactive services will become an essential feature of DTV services from broadcasters, in particular digital cable TV operators, to attract more subscribers and establish a new source of differentiation and income. They will form part of a broadcaster's TV subscription package and reach consumers as a bundled service together with TV programs.

The interactive digital set-top box will dominate the ITV market. By 2003, Europe and the United States will have an installed base of 61 million set-top box of ITV. In the medium- to long-term, increasing competition from large DTV broadcasters will limit the success of dedicated Internet TV services. Their growth will peak in 2002 as the market approaches saturation.

The broadcast television industry can readily participate in a rapidly emerging bandwidth marketplace. Indeed, broadcasters are favored with several inherent competitive advantages, including currently deployed network, wireless distribution, ubiquity in the local market, cost-effectiveness in scale, and the ability to support IP multicasting. We shall discuss the broadcast business model in Section 5.7.

5.2.2 Internet

The Internet is moving fast from its text-based origins beyond graphic enhancements and into streaming audio and video services and full-fledged multimedia. The multimedia market, in the broadest sense, embraces delivery of analog and digital multimedia content via arbitrary packaged and networked means (such as CD-ROMs and DVDs, as well as various telecommunications pathways) to arbitrary devices [TVs, VCRs, set-top boxes, PCs, personal digital assistants (PDAs), and similar devices]. Of course, the Internet (TCP/IP) is the runaway winner for distributing multimedia content in one-to-one and one-to-many connections that are PC-based.

The Internet is going to be the key enabler of Web-based commerce. As per a survey report from Forrester Research, Inc., conducted in 2000, worldwide Internet commerce—both business-to-business (B2B) and business-to-consumer (B2C)—will hit $6.8 trillion in 2004. North America represents a majority of this trade, but its dominance will fade as some Asian-Pacific and Western European countries hit hypergrowth during the next two years. According to another survey by Gartner Group, the global B2B e-commerce market is set to reach $8.5 trillion in 2005. In 2000, the value of worldwide B2B Internet commerce sales transactions exceeded $433 billion, a 189% increase over 1999 sales transactions. Gartner believes that the tight economic climate will precipitate further growth in B2B e-commerce areas such as e-procurement and e-marketplaces [2, 3]. The number of people surfing the Internet while watching TV continues to grow. According to the studies, 8 million people surfed the Internet while watching TV in 1998. That number grew to 27 million people in 1999. Studies further reveal that active on-line users are more than twice as likely as casual users to pay more than $200 for an ITV set-top box. In addition, 60% of consumers who simultaneously access the Internet on their PC and watch TV are interested in ITV set-top boxes, compared with only 47% of users who engage in these activities separately.

5.2.3   E-Commerce

E-commerce is the business transacted through telecommunications where the payment is made electronically. The Web has been the main driver for e-commerce. A user is able to browse a site on the Web, select goods or services, place an order, and pay for items with a credit card. A secured transaction keeps the transaction private. The system is able to authenticate the credit card, check the availability of the balance in the account, debit the account, etc. A number of new innovative ideas are now emerging, such as e-cash and e-check, where the user can purchase these in advance and use them to make payments.

5.2.4   T-Commerce on DBS

T-commerce is e-commerce using the television. We discussed the examples of Wink in Section 3.9.1. Broadcasting to consumers using DBS was discussed in Section 1.6.6. DBS has t-commerce potential. For example, EchoStar and DIRECTV provide satellite TV services to almost 13 million homes in the United States. An agreement to deliver Wink's Enhanced Broadcasting™ feature to EchoStar Communications Corporation (Nasdaq: DISH, DISHP) and its DISH Network™ customers was reached in June 2000. Wink is already available to DirectTV subscribers. This will enable DISH Network subscribers who have OpenTV-enabled set-top boxes by 2001 to receive Wink enhanced programming and advertising and use Wink's Response Network™, described in Section 3.9.1.4. The set-top boxes are predicted to reach 4 million homes by mid-2003.

Satellite users are 50% more likely than cable users to subscribe to ITV services. A study by TechTrends, Inc., reveals that DBS subscribers will pay more than premium cable subscribers for ITV set-top boxes. Among those who are interested, 25% of DBS subscribers will pay more than $200, compared with 20% of premium cable subscribers [4].

5.2.5   TV Portals

Network operators around the world are investing heavily to upgrade their networks to enable new revenue-generating services. Network operators use the TV portal, essentially an interface to interact between customers, service providers, and the Internet, for ITV offerings. The TV portal is analogous to the Internet portal, generating revenue from advertisements and commerce. The TV portal is a branded aggregation of products and services designed to fulfill a large majority of users' needs, such as shopping, banking, news, e-mail, chat, entertainment, VOD, and other interactive services. New business models are being developed to explore the possibility of new revenue streams. The business model does not depend on high subscription fees from ITV subscribers, but rather on low-cost access to the TV portal and earning revenues through advertising, commerce, and tenancy agreements.

TV portals offered by cable, satellite, and telephone network operators generally have walled garden, extended TV services, and enhanced TV broadcasts. These services are integrated in the TV portal platform. Let us examine the opportunities to generate revenue streams. According to Jupiter Communications, 60% of ITV revenues will ultimately be derived from commerce, and 40% will come from advertising. The study does not expect revenue from enhanced subscription fees.

TV portal platforms also enable targeting of customers with enhanced broadcasting of commercials, dynamic advertisements in the portal environment, and direct-response advertising. They allow operators to monitor and analyze broadcasting's effectiveness by using a back-end system that keeps track of clicks on banners and resultant matured orders. Comprehensive reports are made available to network operators and advertisers. A variety of software is available that allows advertisers to pay in terms of actual utility rather than on the basis of a fixed tenancy agreement. This means that an advertiser does not pay for space on the basis of the number of hits being received by a site, but rather on the basis of real business generated by advertising on a specific Web site.

It is clear that network operators can generate significant revenue through their TV portals; however, the success of a TV portal depends on customer satisfaction. Customers need easy access to information via a simple navigation system through a remote control [5].

5.2.6　Walled Garden

Walled garden offers immense potential for ITV commerce. Impulse purchases from direct response commercials and enhanced programming will capture maximum revenue. According to Forrester Research, "…walled gardens will demonstrate where the real power lies—with the operator" [6]. Sharing advertising revenue and tenancy rent will generate this revenue. The model is identical to one being followed by Internet portals. Two examples clarify the fact. FirstUSA Bank paid $500 million to be the exclusive marketer of credit card products on AOL properties for 5 years. Barnesandnoble.com paid $40 million for 4 years as the exclusive bookseller in the

AOL shopping channel [7]. Cable and satellite operators will build walled gardens, generating more than $3 billion in commerce by 2002.

5.2.7   E-Mail

E-mail is a key driver for the success of any ITV technology. One of the main reasons behind the Internet's success is its use for e-mail. According to a recent study, 66% of PC users go on-line to use e-mail. Many people feel that the PC is just an expensive e-mail machine. Interactive digital TV (iDTV) has the opportunity to become the major point of on-line access for the current off-line population. The main requirement is that iDTV companies adopt e-mail-friendly strategies.

5.2.8   Content

Consumers who have the most exposure to TV programming and on-line content exhibit more interest and less price sensitivity than other consumers. These consumers represent substantial opportunity for providers of ITV products and services

For ITV to be successful, the content has to be presented in a TV-centric format. This needs special emphasis on site design. An ITV receiver has low resolution, does not support several standard Web languages, and does not use a mouse navigation interface. Content designers need to better understand these ITV design features. Content for ITV design is much more than providing the ability to surf the Web on a TV. Content design is further complicated by the proliferation of hardware and software platforms. At the same time, this new environment enables the development of many innovative programs and also guarantees a higher level of security to its customers for transactions and information exchange. Network operators need an efficient TV portal solution that can be deployed across multiple technology platforms and network architectures. Open standards are being developed and a number of efforts are ongoing (Section 3.7).

5.2.9   Metadata

Metadata is the information about programs and commercials embedded in video streams. Metadata will allow programs to be embedded with commerce information, videos custom-assembled to viewer preferences, and layered commercials that invite viewers into deeper interactions. Viewers may be able to retrieve players' statistics, get details about a commercial, compare prices, etc. It is expected that by 2003, networks and operators will finally agree on a common standard for metadata. This will unlock the door to new TV behaviors. It is expected that this type of programming will generate $7 billion in subscriptions, $17 billion in marketing fees and advertising, and $23 billion in commerce by 2005 [8].

5.2.10 Metacontent

Metacontent is the content about consumers. Software programs may be used to gather information about consumers, such as watching and surfing habits, preferences, lifestyles, spending habits, etc. In the new convergent marketplace, consumer metacontent is going to be as important as the content itself. The owners of metacontent are going to be extremely powerful in the new convergent marketplace. This raises a number of issues, however, related to privacy (Section 6.3). Laws are also being written to control the power of people collecting personal data.

5.2.11 Data Broadcasting

Direct-to-PC service via satellite has been operating for quite some time. In Europe, this offering has developed as a separate entity than DTV services. In the United States, however, some operators are starting to collaborate with broadcasters. Hughes even offers both types of service—and there are signs that this may happen in Europe. Data service providers also have plans to offer interactive services, including Internet services.

5.2.12 VOD

VOD through DTV is a preferred service. Microsoft's WebTV small-scale trial in the United Kingdom involving 115 households found that 73% wanted VOD services.

5.3 Emerging Markets

Examining ITV opportunities requires a detailed understanding of major business and technology issues. An analysis of the ITV marketplace, covering both iDTV and the dedicated Internet TV, should look at the ITV value chain, service provision issues, and technology issues. The ITV value chain includes content, networks, and interfaces. Service provision issues involve ITV tariff, ITV applications, and partnerships with content providers.

Technology issues involve hardware and software specifications, standardization, and the extent of future support for HTML. We discussed hardware, software, and standardization issues in the earlier chapters. We will now discuss trends and market predictions for ITV. We will use forecasts and survey reports from several companies and agencies. These reports and forecasts do not necessarily match. Further, it cannot be guaranteed that all the forecasts will come true; however, these certainly point to a promising future for ITV, t-commerce, and related technologies.

5.3.1   A Few Trends and Predictions

International telecommunications infrastructures are new and can be upgraded. This has resulted in deployment of ITV by many network operators around the globe. ITV services have been rolled out in several European, Asian, and Latin American countries. Europe is ahead of other countries, with nearly every major network operator currently offering some ITV services to more than 14 million ITV users in Western Europe, compared with less than 1 million users in the United States, according to a survey by Jupiter Communication undertaken in July 2000 [9]. If we consider worldwide statistics, Table 5.1 illustrates the status of a few key enablers.

Interactive services such as e-mail, interactive shopping, banking and Web access via the TV set will be available to one in four households in Europe and the United States by 2003. This will work out to about 67 million households. With TV penetration close to 100% in Western Europe and the United States, the market for ITV is expected to grow at a rate of

Table 5.1
Worldwide Growth of a Few Key Enablers of ITV (U.S.$ millions)

Details	2000	2002	2004
Cable telephony service revenue[1]	801	3,174	7,242
Internet market revenue[2]	39,059	57,881	74,893
Worldwide e-commerce[3]	657,000	2,231,000	6,790,000

1. *After:* Multichannel News International, fact pact, 2001.

2. *After:* PricewaterhouseCoopers.

3. *After:* Forrester Research.

45% over the next 5 years, opening new revenue opportunities for broadcasters and ITV industries. TV-based Web access in Europe and the United States is projected to grow to 10.9 million in 2003. Many new Asian-Pacific and Latin American markets will open. The total user base of ITV services, including DTV and dedicated Internet, will open new revenue streams. We will now examine the worldwide trend of related technology growth that will have an impact on ITV's potential and also discuss growth trends and the absorption of new technologies by various regions.

5.3.1.1 Worldwide E-Commerce Revenue

E-commerce will take place through the Internet, mobile phones, ITV, and other appliances. This will generate massive revenue to be shared by various service competitors. Despite some recent hiccups, the value of global e-commerce markets will surge over the coming years. Worldwide e-commerce growth is given in Figure 5.1.

5.3.1.2 Worldwide Internet Market Revenue

Internet revenue, particularly advertising, will also increase. The worldwide Internet market revenue is given in Figure 5.2. Owing to the likely introduction of a standard API that supports HTML, IDTV service providers will increasingly offer Web access.

There are already a number of offerings available to view the Web via a TV. The world's leading Internet players will merge with complementary vendors in adjacent positions in the industry's value chain. Large dominant

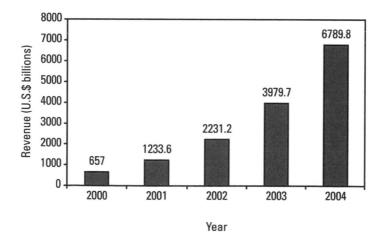

Figure 5.1 Worldwide e-commerce growth. (*Based on data from:* Forrester Research.)

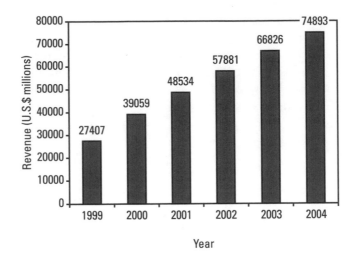

Figure 5.2 Worldwide growth of Internet market revenue. (*Based on data from:* Price-waterhouseCoopers.)

players will emerge after convergence. This process will result in Web content providers delivering their content through other methods.

5.3.1.3 Worldwide ITV Homes

ITV gives broadcasters the chance to compete with Internet service providers for e-commerce revenues. Next-generation advanced set-top boxes will also position TV service providers as key gateway owners in future interactive homes. By 2005, 179 million households worldwide will access a variety of enhanced and interactive on-line services by means of their TV sets [10].

5.3.1.4 Worldwide TV Commerce

Widespread penetration of DTV will lead to convergence between the Internet and traditional TV sector. DTV offers huge potential and rewards for those companies that can implement it in time, provide the right content, and retain their customers. According to Ovum Research, the DTV market can be worth more than $100 billion by 2005. TV subscription services, video gaming, and TV-based information service revenues will exceed $60 billion. T-commerce revenues will reach $45 billion.

5.3.1.5 Worldwide Smart TV Revenue

Advanced TV devices and interactive content will dramatically change how millions of viewers consume TV programming. Television is about to get a whole lot smarter. ITV will penetrate rapidly in households. According to Forrester Research, by 2005, there will be 87 million DBS customers with Interactive Programming Guide (IPG), 65 million households that can interact with video, and 53 million PVR users. New revenue streams generated by interactive and personal TV are expected to reach $7 billion in subscriptions, $17 billion in marketing fees and advertising, and $23 billion in commerce. Even as they drain $18 billion in ordinary TV advertising revenues, smarter devices will create $25 billion in new revenues from viewers interacting with their TV screens [11].

5.3.1.6 Revenue Through Non-PC Devices

Internet-based applications typically presume a PC client. This is changing with the introduction of non-PC clients becoming known as a class of "Internet devices." This includes PDAs, cellular and PCS phones, and an increasing array of consumer electronics devices. ITV and other non-PC devices will have $23 billion in on-line sales in 2005, and influence another $128 billion in off-line sales [12].

5.3.1.7 Personal TV Services

Personal TV services will reach widespread acceptance through multipurpose set-top boxes rather than single-purpose DVR devices, as previously expected.

5.3.2 Regional Growth Potential

In this section, we will examine trends, projected growth, and the potential of various connected technologies in Europe, North America, Asia-Pacific and other places. However, before discussing growth potentials, we will define the terms used in connection with TV and Internet-related business.

E-commerce, e-commerce, ecommerce, or electronic commerce. E-commerce is defined as the conduct of a financial transaction by electronic means. With the growth of commerce on the Internet and the Web, *ecommerce* often refers to purchases from on-line stores on the Web, otherwise known as *e-commerce Web sites.* They are also referred to as *virtual-stores* or *cyber stores.* Since the transaction goes through the Internet and the Web, some have suggested another term: *I-commerce* (Internet commerce), or *icommerce.* Few have referred to it as *Web-commerce.* However, e-commerce is not limited to the Web only and includes all financial transaction by electronic means. E-commerce can be B2B or B2C.

T-commerce or t-commerce.   T-commerce provides e-commerce opportunities via the most ubiquitous of all mediums, the television.

Internet Market Revenue.   Internet market revenue is the total revenue derived in terms of retail subscriber base and dollar value sizing of the ISP market to all players [including ISPs, network operators, content providers, application service providers (ASPs), equipment vendors, and other service providers].

Total Advertisement.   This is the revenue to be earned by all the media through advertisement of space or slot on the medium.

TV Advertisement   The revenue earned by advertisement and sale of spots on TV medium. This is a segment of total advertisement revenue.

Internet Advertisement   The revenue earned by advertisement and sale of spots or banners on Internet sites. This is a segment of total advertisement revenue.

On-Line Revenue.   This is the revenue earned through electronic payment via payment gateways using credit cards, e-check, cyber cash, etc.

Difference Between Internet Market Revenue and E-Commerce.   Internet market revenue is the total revenue earned through the Internet. This includes ISPs, content providers, ASPs, equipment manufacturers, and other players. E-commerce is only one portion of the Internet's market revenue, which is earned through financial transactions by electronic means. E-commerce also includes revenue earned by electronic means using other networks.

5.3.2.1   Europe

The European market is home to a number of iDTV platforms. A summary of the appliances available in European households is given in Table 5.2. It can be seen that almost every household has access to a TV set.

Only 27.7% of European households have a PC; However, the computer is a dominant platform for interactive services. It is predicted that computers will continue to be the main device for interactive services in the European market in the foreseeable future. By the end of 2002, there will be 40 million on-line-enabled PC households in Europe, compared with 15 million households with digital set-top boxes (Figure 5.3). The reason for this is the sluggish development of DTV offerings in Europe. Countries are

Table 5.2

Household Appliances—Percentage of European Homes

Household Appliance	Percentage of European Homes
Telephone	90.7
Television	98.2
Cable television	43.4
Satellite television	13.7
Video recorder	61.1
Personal computer	27.7
Video games	17.2

From: Ethos, 2000.

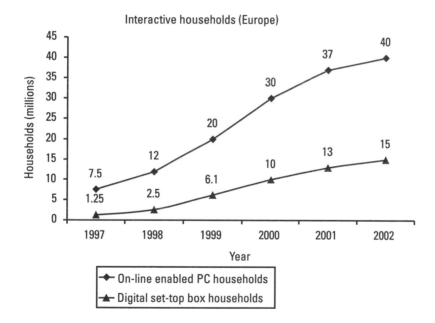

Figure 5.3 Interactive services via computers versus set-top boxes. (*Based on data from:* Datamonitor.)

reluctant to announce a firm date for switching off analog systems due to the uncertainty about the progress of digital television and the value of the spectrum. Nevertheless, many countries are targeting 2010 to terminate analog systems. DTV is expected to reach 80 million subscribers in Europe by 2005, when more Europeans will use ITV than go on-line with PCs. Further, DTV is especially big in such countries as the United Kingdom and France, where viewers get only three to five broadcast channels without a subscription.

The European on-line market revenue due to advertisement is projected to grow from $450 million in 1999 to $6 billion in 2004. The growth is given in Figure 5.4. The European on-line access revenue is projected to jump from $9.07 billion to $18.6 billion during the same period (Figure 5.5). The total European on-line revenue is shown in Figure 5.6. A few other indicators for Europe are described below:

- Forty-three percent of consumers would pay at least $3 per month for VOD.

- More than 50% of European households will have some form of interactive DTV service by 2005.

- By 2005, there will be 87 million households that use interactive program guides, 65 million households that can interact with video, and 53 million PVR users.

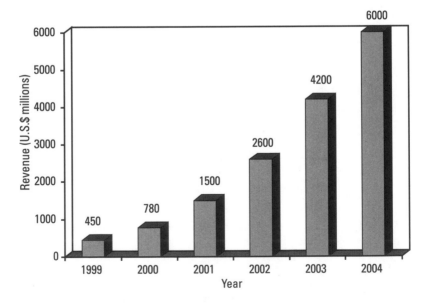

Figure 5.4 European on-line advertisement revenue. (*Based on data from:* Pricewater-houseCoopers.)

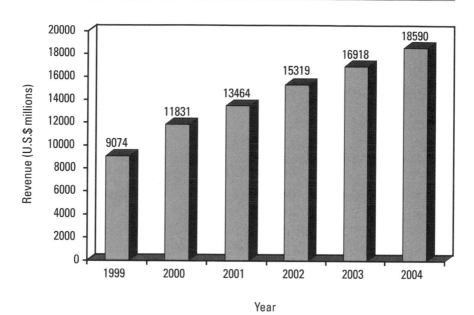

Figure 5.5 European on-line access revenue. (*Based on data from:* Pricewaterhouse-Coopers.)

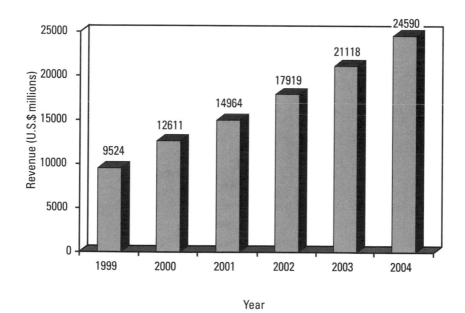

Figure 5.6 European on-line total revenue. (*Based on data from:* PricewaterhouseCoopers.)

The e-commerce growth in Western Europe is given in Figure 5.7. It can be seen that the revenue will grow from $87.4 billion in 2000 to $1.53 trillion in 2004. This revenue may be captured by t-commerce. Total advertising revenue and advertising revenue shared by TV is given in Figure 5.8.

In Europe, 20 million households receive ITV through satellite services. A survey of households across Great Britain, France, Germany, the Netherlands, and Sweden, reveals that 50% of European households will have some form of iDTV service by 2005. Another projection suggests that by 2005, there will be nearly 70 million "convergent consumers" in Europe, using all three key digital platforms: DTV, mobile or wireless, and the Internet. (This figure excludes customers using only one or two of the three platforms.) It is expected that t-commerce will be worth $28 billion by 2005 and account for about 16% of all on-line retail sales. Forrester Research predicts that 80% of all European households will have ITV by 2010. ITV will be the key technology to bring e-mail and other on-line interactive services into homes that do not yet have Internet access. ITV service operators currently delivering service using various delivery modes in Europe are described in Table 5.3.

Europe is ahead of the United States when it comes to interactive TV. Canal Satellite has 14 million European subscribers, and 96% currently access interactive applications at least once daily. Canal+ is an ITV service provider that primarily operates in Europe, and Canal Satellite is one of its

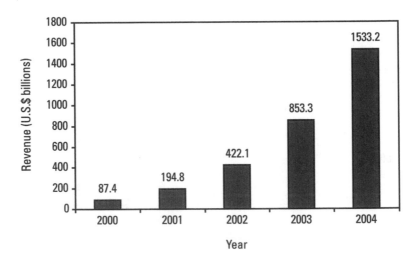

Figure 5.7 E-commerce growth in Western Europe. (*Based on data from:* Forrester Research.)

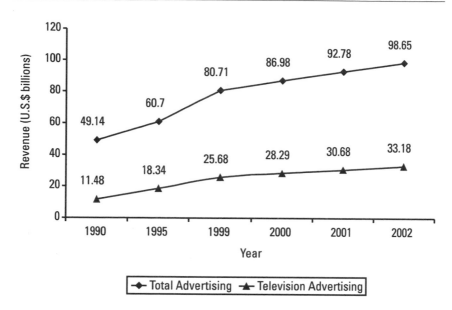

Figure 5.8 Total advertisement vis-à-vis TV advertisement revenue in Europe. (*Based on data from:* Forrester Research.)

divisions. Scandinavia and the United Kingdom have the highest penetration of cross-platform consumers of DTV, Internet, and mobile communications and the best overall infrastructures across the three platforms, with Belgium and Italy the worst. Projections show that France, Spain, and Italy will be TV-focused, with more ITV households than PC ones. The PC as a consumer interactive device will dominate in the German-speaking markets. The United Kingdom, along with Sweden, will show a high penetration of both interactive technologies.

Interactive services will reach more than 19% of households in the United Kingdom, 28% in Sweden, and 12% in France by 2002. Approximately 33% of U.K. households and 29% of French households are willing to pay for interactivity on their TV sets; iDTV homes are projected to increase rapidly (Table 5.4).

At present nearly half of European households have cable TV (both analog and digital). By 2003, there will be 7.2 million cable, 17.4 million satellite, and 5.1 million terrestrial digital set-top boxes in Europe. Growth will be strongest in the United Kingdom, France, Italy, and Spain, where digital pay-TV operators such as BSkyB, Canal Satellite, Tele+, and Via Digital have content packages for viewers to choose from. Digital services are rolling out steadily on cable networks, while digital terrestrial launches are lagging

Table 5.3
ITV Service Operators in Europe

Country	Operator	Delivery Mode	Launch date	API
Denmark	Tele Danmark	Cable	1999	OpenTV
France	Canal Satellite	Satellite	1997	MediaHighway
	TPS	Satellite	1997	OpenTV
	Lyonnaise Cable	Cable	1997	OpenTV
Germany	PremiereWorld	Satellite, cable	2000	Betanova
Italy	Telepiu	Satellite	2000	MediaHighway
	Stream	Satellite	1999	OpenTV
Nordic	Canal Digital	Satellite	1999	MediaHighway
Spain	Canal Satellite	Satellite	1997	MediaHighway
	Via Digital	Satellite	1999	OpenTV
	Onda Digital	Terrestrial	2000	OpenTV
Sweden	Telia	Cable	1998	OpenTV
	Senda	Terrestrial	1999	OpenTV
United Kingdom	BSkyB	Satellite	1999	OpenTV
	CWC*	Cable	1999	Liberate TV Navigator
	NTL	Cable	1999	PowerTV
	ONDigital	Digital Terrestrial	2000	MediaHighway/MHEG5
	Telewest	Cable	2000	Liberate TV Navigator

*CWC (Cable and Wireless Communications) was acquired by NTL

Table 5.4
iDTV Homes in Europe (millions)

Service	1999	2000	2002	2005
Walled garden	6.9	12.8	20.3	15.1
Upgrades to unwalled services	0.0	00.2	07.0	31.7
New unwalled services	0.1	01.0	05.0	23.0
Free-to-air iDTV services	0.0	00.0	01.0	10.0

From: Forrester Research.

behind. Europe had 362,000 broadband households in 1999, which is projected to grow to 28.7 million by 2005.

United Kingdom

Over 60% of households surveyed by the U.K. Office of Telecommunications in July 2000 claimed to have interactive services such as home shopping, and 34% claimed to have e-mail and Internet access via the TV. In the United Kingdom, digital satellite TV provider British Sky Broadcasting Group provides Open interactive service for free. ONdigital broadcasting service in the United Kingdom has 550,000 homes using the Canal+ systems for free and pay ITV services. Cable & Wireless in the United Kingdom reports 60,000 digital cable customers using its Liberate ITV system. The SkyDigital satellite service in the United Kingdom now has 2.6 million customers, half of who use the OpenTV ITV service. In Western Europe, Canal+ claims 4 million ITV subscribers among 13 million digital satellite and cable customers. The e-commerce revenue in Western Europe will grow from $87.4 billion in 2000 to $1.5 trillion in 2004.

France

The French-language digital DTH service run by GlobeCast and the Office des Postes et Télécommunications of French Polynesia consists of multimedia and Internet services. Canal Satellite has been offering interactive services to 1.5 million subscribers for nearly four years. Banking is one of 30 available applications. Canal subscribers have been able to use their chip-based *Cartes Bancaires* charge cards from the beginning to buy pay-per-view programs and do on-screen shopping. The operator has signed up 18 merchants and offers TV banking and gambling. The service is provided in association with PMU, Europe's largest *parimutuel* company. The betting accounts are operated using a chip-based payment card issued by French financial institutions. Betters can transfer up to 600 French francs (U.S.$85) by inserting their chip-based payment cards into a slot on their set-top boxes. About 8,000 viewers opened horse race betting accounts within the first two weeks of the service's launch. The betting channel got more than 15,000 visits per day in the two weeks following the launch.

5.3.2.2 United States and North America

In the United States, cable is a popular delivery system for digital programs and ITV. AT&T, Time Warner, and Comcast, the largest cable operators, had a combined 2.7 million subscribers wired for digital programming at the end of 1999. In the United States, the number of digital-video subscribers reached 5 million in 1999. This is expected to increase to 33.2 million in 2006 as per Canadian Imperial Bank of Commerce (CIBC) forecasts (www.cibc.com). Cable operators are upgrading their networks to provide two-way interactive service. The top 10 multiple system operators (MSOs), reaching 75% of cable subscribers, had planned to upgrade more than 80% of their plants to two-way interactive system by the end of 2000. By 2004, virtually every U.S. cable TV plant will be capable of offering two-way inter-active services [13]. With the completion of these upgrades, the ITV industry is poised for explosive growth. By 2005, the U.S. ITV market will generate approximately $25 billion in annual revenue [14]. This represents a huge market opportunity for a wide variety of companies, including cable, satel-lite, and telephone network operators, technology providers, and e-business companies. Cable is expected to be the dominant DTV delivery system in the United States.

In the United States, satellite providers have dominated the DTV mar-ket. DirecTV and EchoStar together claim more than 11 million subscribers. All 11 million of the satellite decoder boxes contain conditional access smart cards to authenticate subscribers to receive the service.

Other indicators also promise fast ITV growth. E-commerce growth in North America (the United States, Canada, and Mexico) is given in Table 5.5 and Figure 5.9. Internet revenue growth in the United States is given in Figure 5.10. Total advertising revenue and TV advertising revenue in North America are shown in Figure 5.11.

The cost of a set-top box has fallen from $4,000 to a couple hundred dollars. There are 5 million subscribers to ITV services in the United States. The number will increase to 65 million by 2005. By the end of 2005, ITV

Table 5.5
E-commerce Growth in North America (U.S.$ billions)

Year	Revenue in United States	Revenue in Canada	Revenue in Mexico	Total
2000	488.7	17.4	3.2	509.3
2001	864.1	38.6	6.6	909.3
2002	1,411.3	68.0	15.9	1,495.2
2003	2,817.2	109.6	42.3	2,969.1
2004	3,189.0	160.3	107.0	3,456.3

From: Forrester Research

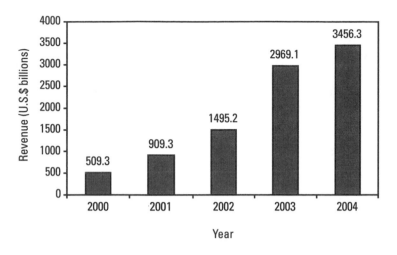

Figure 5.9 E-commerce growth in North America. (*Based on data from:* Forrester Research.)

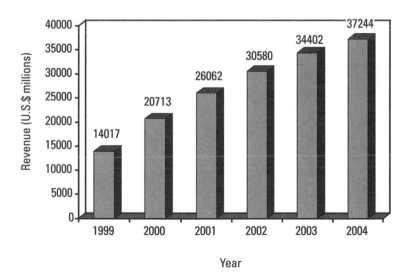

Figure 5.10 Internet market growth in the United States. (*Based on data from:* PricewaterhouseCoopers.)

services in the United States will generate $7 billion in subscriptions, $17 billion in marketing and advertising fees, and $23 billion in e-commerce.

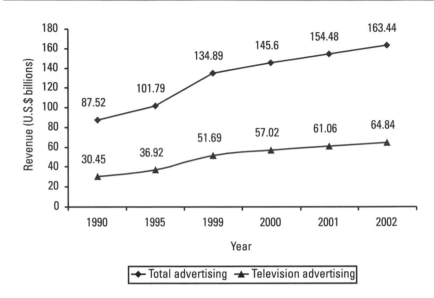

Figure 5.11 Total advertisement revenue vis-à-vis TV revenue in North America. (*Based on data from:* Zenith Media.)

5.3.2.3 Canada

ITV is available in Canada to a small population only. The service is provided by Microsoft Corp.'s WebTV. It is largely used for e-mail access by those who cannot or do not want to use a PC. The service has been limited by slow dial-up access to the Internet. Faster broadband cable and digital-subscriber-line phone connections are now available. ITV services are being launched by Internet portal AOL Canada Services, Inc. in Toronto and cable companies such as Toronto-based Rogers Cablesystems Ltd. and Montreal-based Groupe Vidéotron Ltée. Other companies, such as Shaw Communications, Inc. in Calgary, BCE, Inc. in Montreal and Burnaby, B.C.–based Telus Corp., are working on their versions of ITV. These services will use digital set-top boxes that are already available to subscribers of digital cable TV. As per forecasts of the Canadian Cable Television Association, there were 5.3 million digital-capable households in Canada in January 2000, which is expected to increase to 6.9 million in 2005.

5.3.2.4 Latin America

Latin America had just 2.65 million multichannel homes in 1990. This grew to 19 million in 2000. It is projected to increase to 27.6 million by 2004. Latin America's TV advertisement expenditure was $2.91 billion in 1990, which grew to $11.62 billion in 2000 (Figure 5.12). The region's pay-TV subscription revenue was U.S.$ 6.6 billion in 2000. Latin American had 9.3 million on-line users in 1999. This is expected to grow to 37.6 million in 2003 (Jupiter Media forecast). On-line advertisement spending grew by 111% from 1999 to 2000. E-commerce business reached $3.6 billion in 2000 and is expected to grow to $31.8 billion by 2003, according to Forrester Research (Figure 5.13). The region promises good growth for ITV services in coming years.

5.3.2.5 Asia-Pacific

A study titled "Riding the Next Wave: The Strategic Implications of Interactive TV and Broadband Services in Asia Pacific" by Andersen Consulting found that DTV represents an enormous and growing opportunity in the Asia-Pacific region. This can put the region at the forefront of the development of ITV and its use for e-commerce. The study included China, Japan, Hong Kong, Taiwan, Australia, South Korea, Singapore, and India.

A number of factors contribute to the area's high potential, including higher rates of TV penetration than any other region in the world, and a substantial and growing broadband subscriber base already in place in some markets. The region has high TV penetration, extensive cable TV networks, strong growth in broadband Internet penetration, and a youthful population. The

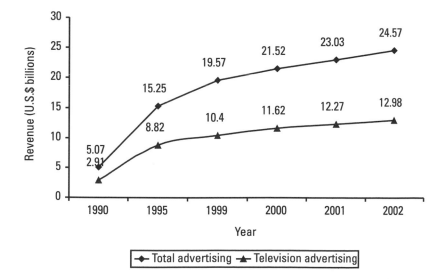

Figure 5.12 Total advertisement vis-à-vis TV advertisement revenue in Latin America. (*Based on data from:* Zenith Media.)

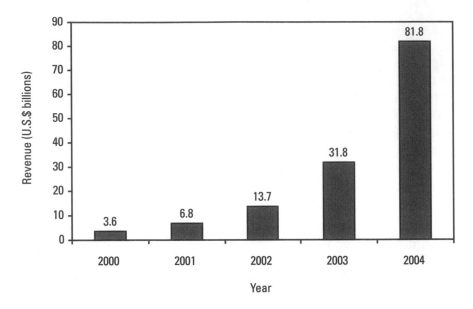

Figure 5.13 E-commerce growth in Latin America. (*Based on data from:* Forrester Research.)

region has 470 million TV households. In China's major cities, cable TV penetration is close to 100%. In India, there are about 30 million cable households. All major Asian countries have acquired domestic satellites to meet the rising demand for increased bandwidth for corporate data, multimedia, and Internet services. Many countries in the region are planning substantial DTV services in the next 2 to 3 years. Governments across the region are actively encouraging the development of DTV by deregulating broadcasting and tele-communications industries and opening markets to competition.

Let us examine a few key technologies and their growth. E-commerce growth in Asia-Pacific is given in Figure 5.14. Internet revenue growth is given in Figure 5.15. Total advertising revenue and TV advertising revenue are shown in Figure 5.16.

The Asia-Pacific on-line market revenue due to advertisement is projected to grow from $245 million in 1999 to $3,000 million in 2004. The growth is given in Figure 5.17. During the same period, on-line access revenue is projected to jump from $3,621 million to $10,059 million (Figure 5.18). The total on-line revenue is shown in Figure 5.19. A few of the other indicators for the Asia-Pacific region are described below:

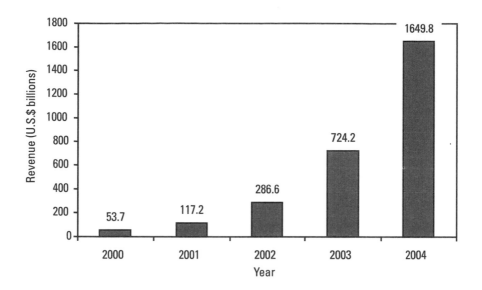

Figure 5.14 E-commerce growth in Asia-Pacific. (*Based on data from*: Forrester Research; includes Japan, Australia, and other countries.)

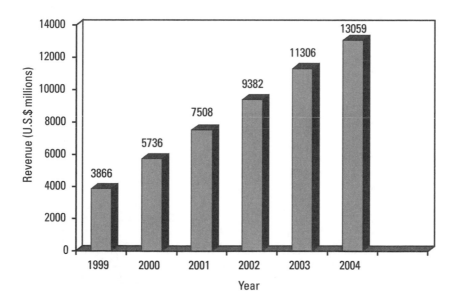

Figure 5.15 Worldwide Internet revenue growth. (*Based on data from*: Pricewater-houseCoopers; includes Japan, Australia, and other countries.)

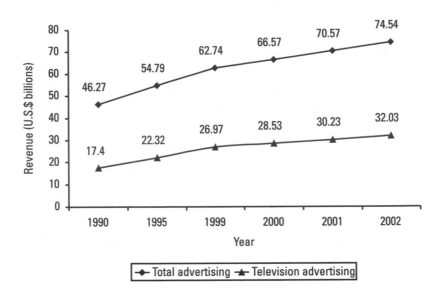

Figure 5.16 Total advertising revenue vis-à-vis TV advertising revenue. (*Based on data from:* Zenith Media; includes 14 biggest TV markets in Asia, including New Zealand and Australia.)

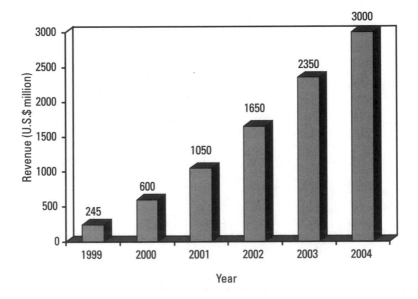

Figure 5.17 On-line market revenue due to advertisement in the Asia-Pacific region. (*Based on data from:* PricewaterhouseCoopers.)

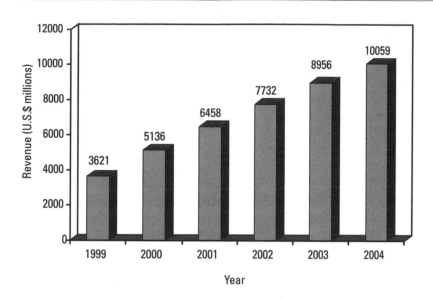

Figure 5.18 On-line accesses revenue in the Asia-Pacific region. (*Based on data from:* PricewaterhouseCoopers.)

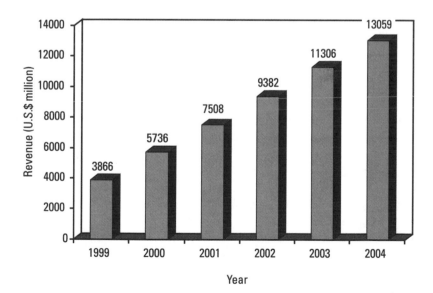

Figure 5.19 Total Web revenue in the Asia-Pacific region. (*Based on data from:* PricewaterhouseCoopers.)

- Total cable households to increase from 144.69 million in 2000 to 198.02 million in 2004 [15];

- Total satellite households to increase from 22.10 million in 2000 to 29.63 million in 2004 [15];

- Television advertising expenditures to increase from U.S.$25.73 billion in 2000 to U.S.$34.36 billion in 2004 [15];

- Basic services subscription revenue to grow from U.S.$25.63 billion in 2000 to U.S.$39.39 billion in 2004 [15];

- Premium services subscription revenue to grow from U.S.$1.78 billion in 2000 to U.S.$6.67 billion in 2004 [15].

China

With a population of 1.3 billion, China currently has about 30 million TV sets, 40 million video compact disc (VCD) players, 11 million PCs, and 22.5 million Web surfers. However, there were 35.6 million e-mail accounts as of December 1999. China's TV advertising expenditure was U.S.$1.9 billion in 1999 and is poised to grow to U.S.$4.2 billion in 2004. That' is a huge market for growth. Microsoft is now a major player in China's convergence activities with its "Venus" project. Open TV, a U.S.-based ITV software company, has made a deal with the China Broadcasting Film and Satellite Company (CBSat) to become the first Chinese mainline provider of ITV. Optimistically, it aims to provide 2.5 million set-top boxes by 2006. The untapped Chinese market, a mixture of communism and capitalism, is one of the world's most attractive markets. ITV may appeal to the new wealth created by the urban population. Open TV has successfully deployed its technology in many countries and plans to expand in China's market in a big way.

ITV needs infrastructure that provides enough bandwidth and credit cards to allow consumers to buy products off the TV. Visa International expects to issue 100 million credit cards in China by the end of 2001. The Chinese system will be satellite based due to the high expense of providing potentially millions of miles of cable to cover the country.

India

India is a vast country with one billion people. It had 100 million radio receivers, 80 million TV receivers, 30 million cable subscribers, and 2.7 million satellite households as of December 2000. India sells more than 6 million TV sets per annum (3 million color and 3 million black and white). Penetration of cable TV in India has increased from 412,000 households in

1992 to 30 million households in 2000 (estimates vary between 25 and 35 million households). Assuming a 15% growth per annum, cable TV penetration would touch 50 million households by 2003. Television advertising expenditure was U.S.$576 million in 1999. India's basic subscription revenue was U.S.$14.6 billion in 2000 and is projected to grow to U.S.$20.1 billion by 2004. The premium subscription revenue was $314 million in 2000, expected to be $623 million in 2004. The demand for TVs, especially Internet-based WebTVs, is projected to go up with the government deregulating DTH in 2000. Star TV, national broadcaster Prasar Bharati, network operator VSNL, and many others plan to start DTH service. With DTH in place, consumers will be able to receive more than 100 channels with the installation of a single dish, which may cost between Rs 12,000–15,000 (about U.S.$300). This shows the potential of the Indian TV market. Knowing this, consumer durable majors like BPL, Videocon, MIRC, and Samsung have already launched WebTV in India. In fact, Samsung has gone a step ahead by launching set-top boxes that enable consumers to surf the Internet using a conventional TV. Media giants in India like Zee, Sun Network, and In-cable Network have big plans for the Internet through cable (broadband). Therefore, bandwidth will not be a problem. This would enable consumers to buy or sell through television via t-commerce. Given the fact that computers are not affordable to many, Web-based TV holds great promise.

Japan

Japan is the most promising country for ITV services. It offers to provide many types of interactive services through its integrated services TV (ISTV), which is a household receiver for integrated services digital broadcasting (ISDB) [16]. ISTV is a multimedia-type TV with a built-in home server. It has three major characteristics:

- Its built-in server would enable viewers to instantly call up news and other information, which would be constantly updated automatically through home interactivity.
- Its media fusion capability would enable joint use of TV with other media.
- Its learning ability would make TV more intelligent.

Japan had 24.5 million cable and satellite households in 2000. Japan's e-commerce business reached U.S.$31.9 billion in 2000 and is projected to grow to U.S.$880.3 billion in 2004. Japan's basic subscription spending was

U.S.$5.6 billion in 1999, which is likely to grow to U.S.$8.6 billion by 2004. Japan's premium subscription spending accounted for U.S.$803 million in 1999.

Australia

Australia had 1.25 million cable and satellite households in 2000. It is projected to grow to 1.90 million by 2004. Its e-commerce business is projected to grow from U.S.$5.6 billion in 2000 to U.S.$207.6 billion by 2004. This shows a staggering growth in the area of e-commerce. Basic subscription spending was U.S.$443 million in 2000, which is expected to grow to U.S.$804 million by 2004. Premium subscription spending was U.S.$188 million in 2000 and is expected to grow to U.S.$409 million in 2004. ITV has already started in Australia. MasterCard International started the first Australian t-commerce transaction via a credit card as part of ICE Interactive's ITV trial in Orange in February 2001. The service enables ITV users to purchase goods and services with their MasterCards from participating merchants at www.mastercardexclusives.com.au. To use the new service, the user clicks through to the MasterCard Exclusives page from the banner advertisement, TV commercial prompt, or shopping ichannel. After selecting from the offers available, the user enters his MasterCard number and other details for processing by MasterCard Exclusives. Once the transaction is processed, a confirmation e-mail is sent to the user and the goods are delivered to the customer.

5. 4 Market Dynamics

The ITV market will grow. However, this will depend on a number of factors discussed earlier in Section 5.2. The growth of ITV may also result in the loss of some traditional markets. As a result, the risks will need to be spread. Smarter TV will enable users to skip commercials, compelling content providers to devise innovative commercials, possibly embedded in main programs as discussed in Section 4.2.2. Collaborations, tie-ups, and mergers will also be required, since a number of players have to work together. It will be profitable to tie-up or acquire the expertise of established operators in key fields rather than developing in-house expertise.

5.4.1 Spreading the Risk

ITV service providers have to create new markets and a critical mass of early adopters before they can create a secure foothold in the market. Therefore, key players are placing stakes in cross-markets to spread the risk. For example, cable companies are also entering digital terrestrial markets. Terrestrial broadcasters, including public sector broadcasters, are entering satellite and cable development fields. Telecommunications and IT hardware companies have made investments in digital broadcasting developments. Software companies are investing in digital broadcasting developments. Many companies are becoming integrated communications companies with the convergence between the TV and PC.

5.4.2 Commercials

Smarter TV will enable viewers to skip 30% of commercials in 2005, causing an unprecedented decline in traditional advertising revenues. This will require new methods of presenting commercials. Targeted networks will embed commerce in programming while real-time information providers will embed codes in their content so that those smart devices are able to reassemble and manipulate it. Broadcast networks will use new models of service and content delivery. Advertisers will need to target individuals based on the viewing data collected by smarter TV devices. Companies wedded to the metrics of plain old TV such as audience sizes and prime-time schedules will be deprived of these new revenue streams.

5.5 Market and Collaborations

The development of ITV holds promise to create vast opportunities for a variety of companies. Explosive growth of t-commerce revenues have been projected after 2003, when first-generation ITV services are expected to attain mass-market penetration. *E-Commerce Times* predicts that the value of B2C t-commerce transactions in the United States alone will grow from $15 billion in 2002 to $105 billion in 2006, overtaking revenues from PC-based e-commerce. Forrester Research estimates TV-based commerce revenues in 2004 to be $1.1 billion from EPGs, $3.8 billion from enhanced television broadcasts, and an additional $2 billion generated from Internet browsing on TV.

The need for a number of players to provide network, content, commercials, and administering ITV billing and customer care services has led to many strategic partnerships. Such partnerships and equity agreements between ITV start-ups and well-established TV and media industry giants expand distribution channels for advanced TV services and create lucrative

revenue-sharing opportunities for the parties involved. Cable, satellite operators, and broadcasters will be the primary beneficiaries of the universal
deployment of ITV technologies and will receive a substantial share of subscription, advertisement, and commerce fees in exchange for distribution and
product marketing support.

The market for interactive and enhanced TV is certainly significant.
There are more than 1.4 billion analog TVs currently in use worldwide. In
the United States, a color TV set is one of the most highly demanded consumer electronics devices and enjoys more than 98% household penetration.
It comes as no surprise that many players in the unfolding ITV industry have
been forming alliances and making strategic acquisitions to gain a better
foothold in what promises to be a heated race for the sector's dominant
position.

5.5.1   Wink

Wink's collaboration with OpenTV, Microsoft, Liberate, PowerTV, Source
Media, and Canal+ Technologies means that Wink's impulse t-commerce
services will now be available on every major software platform currently
offered to U.S. operators. Wink's enhanced programming service is also
available to 200,000 cable households across the United States. The company has distribution deals with the top five cable operators and content
deals with broadcast and cable networks that, when combined, deliver more
than 90% of all TV viewing in the United States. The company also has
deals with more than 20 national TV advertisers. Currently, Wink claims to
be enhancing about 1,200 program hours per week. The service is available
in eight cable systems, reaching 147,000 homes in California, Connecticut,
Illinois, Missouri, New York, Tennessee, and Texas. There is no charge for
subscribers, since the costs are borne by the cable operators and networks
involved. Additional support for Wink comes from set-top box manufacturers (such as General Instrument and Scientific Atlanta), TV set manufacturers (Thomson/RCA, Toshiba, and Matsushita), and TV networks (ABC,
NBC, CBS, Fox, CNN, ESPN, Nickelodeon, TBS, TNT, Lifetime, Food
Network, and USA).

5.5.2   DIRECTV

DIRECTV, and most of the cable MSOs, use Wink's Enhanced Broadcasting system to implement low-cost e-commerce on TV. Microsoft has taken a
10% stake in the company, using Wink as the processing system for its

WebTV devices. DIRECTV has a 5% stake in providing service to two million of its satellite homes. Advertisers on Wink can interact in various ways. They can question viewers directly ("Do you use Brand A?" "What are the most important features of Product B?"). Viewers respond using the remote control keypad. Coupons can be offered ("Click here to receive a coupon"), along with informational brochures. The viewer can also sign up for a test drive with the local car dealer. All responses are first sent to Wink and then on to the advertiser or fulfillment center.

5.5.3 Liberate TV

Liberate TV has collaborated with MediaOne, Comcast, Rogers Communications, US West, Shaw Communications, and America Online. These companies use Liberate TV Navigator and Liberate Connect software to deploy and manage interactive services to their customers.

Liberate Technologies acquired privately held MoreCom, Inc., which provides high-speed Internet-over-TV service via a digital cable TV set-top box and a standard analog TV set. As part of the agreement, Liberate will obtain MoreCom's patent portfolio, which addresses various technologies for merging DTV with Internet content. The acquisition will also significantly expand Liberate's international customer reach, giving the company access to 4 million cable subscribers in Germany and more than 600,000 in Israel. Liberate has also partnered with personal TV company TiVo, Inc. to develop an integrated suite of products for the AOL TV service.

5.5.4 OpenTV

OpenTV is supported by investments from eight leading Internet and broadcasting companies, including America Online, Liberty Digital, Time Warner, and News Corporation. The leading enabler of ITV services, OpenTV has already deployed its software to more than 10 million digital set-top boxes worldwide. Approximately 34 digital cable, satellite, and terrestrial communications networks in more than 50 countries have selected OpenTV solutions to deliver interactive services to their customers. This includes British Sky Broadcasting (BSkyB) in the United Kingdom; TPS and Cable Lyonnaise in France; Via Digital in Spain; Senda in Sweden; Stream in Italy; Galaxy Latin America, the exclusive provider of DIRECTV in Latin America; and EchoStar's DISH Network in the United States. In addition, 29 digital set-top box manufacturers have licensed OpenTV's software, and

OpenTV's authoring tools are licensed to more than 600 independent developers and content service providers.

OpenTV has successfully integrated its digital ITV operating system to the Motorola DCT-2000 set-top terminal, which is currently used in more than 5 million North American homes. Cable operators will be able to download OpenTV's software to already-deployed DCT-2000 terminals anywhere in the United States.

OpenTV has also formed a joint venture with satellite services company EchoStar Communications to develop an OpenTV-enabled, DTV set-top receiver with hard drive. The company is acquiring Spyglass, Inc., which pioneered a number of advanced technologies for Internet-enabled set-top boxes, mobile phones, and other non-PC mobile devices.

5.5.5 Microsoft

Microsoft believes ITV will be one of the company's largest growth areas. It acquired Israel-based Peach Networks to extend the Microsoft TV platform solution to a greater number of cable network operators worldwide. Microsoft has also formed an alliance with NDS, a news corporation company. This partnership will allow Microsoft's Enhanced TV products to reach a larger number of cable, satellite, and terrestrial network subscribers. Microsoft also announced plans to collaborate with Culver City, California–based Intertainer, Inc., which is developing enabling technologies for a broad variety of on-demand entertainment services.

5.6 The Consumer

TV viewers have always responded, directly or indirectly, to what they view. There are many examples of viewers responding in large numbers to interactive telephone response polls in news programming and after major political speeches. In addition, many viewers' channel-changing behavior may be considered a simple form of interacting with TV.

5.6.1 Consumer Preferences

Nonetheless, the basic question of mass consumer appetite for ITV remains unanswered. The behavior of various demographic groups has not been examined. Studies relating responses to a consumer's age, sex, educational background, interest and other parameters have yet to be conducted in a conclusive way. Past experience reveals that certain programs, such as videotext and video games, are preferred by male viewers. Will ITV use skew toward younger audiences? A related question concerns the ability to maneuver through interactive systems. Viewers with experience using PCs and automated teller machines may find ITV easier to use than those who lack familiarity with interactive technologies. What will happen to uneducated people? The education rate in many Asian-Pacific countries is just about 50%. Will programming need to be done in local languages? What happens to dialects and scripts? These are many of the questions that need to be addressed.

5.6.2 Paying Capacity

How many consumers will be willing to pay for ITV services and how much? Bell Atlantic has indicated that it expects consumers to pay $55 per month for video services [17]. Southwest Bell expects combined spending for cable and telephone service to be in the $100–$110 per month range [18]. These may be realistic estimates for consumers in affluent countries but may turn out to be completely unrealistic for many Asian-Pacific and Latin American countries. The average household with cable TV, VCR, and telephone service in the United States spends approximately $110 per month on all three of these services combined, and some households spend considerably more. In many developing countries, the expenditure on these services is $10, even in urban cities.

5.6.3 Entertainment Spending

However, some encouraging news for ITV system builders and service providers comes in the form of trend data about household spending. For decades, household expenditures on entertainment and recreation were a relatively constant proportion of the household budget. Beginning in the 1970s and continuing through the 1990s, the percentage of spending on recreation has increased. With the use of credit cards, availability of more forms of entertainment, and the information explosion, people around the globe are spending more on leisure and entertainment. It is not clear if the share of the household budget on recreation and entertainment will continue to expand or remain level. T-commerce goes far beyond entertainment—most of the spending here would come from other areas of the consumer's budget.

5.7 Emerging Business Model for Broadcasters

The launch of the terrestrial digital television in a few parts of the globe has given broadcasters a new lease on life. There is plenty of excitement and concerns about the prospects of life in the digital lane. In the short run, the significant advantage of DTV is in multicasting. DTV will be able to provide hundreds of channels due to the availability of compression techniques. However, DTV's strategic influence will be felt far into the future in more significant ways than resolution and aspect ratio alone.

Future growth prospects are about new business models. Early interest is fueled by generations of new revenue streams through use of conditional access. Revenue mix can be achieved by also using the service to provide data broadcasts. However, in the long run, additional revenue streams will emerge from distributing other enhanced data and multimedia data types using the ability of DTV to push a different type of data packets. Broadcasters can help address the bandwidth congestion inherent to the rapid acceptance and use of Internet-based applications. Streaming and multicasting are fast-growing applications on the Internet. As these applications scale in the marketplace, inter-network infrastructure of supporting bit streams that are both highly bandwidth-intensive and highly redundant will be under strain. Broadcasters have started recognizing that the systems they build and operate are actually part of a larger information and telecommunications market. Their ability to distribute content beyond traditional services has far more utility and economic value than they may have anticipated.

5.7.1 In Search of New Applications

Broadcasters can use the digital lane for many innovative and killer applications. Let us consider the following applications and scenarios:

- If IP over VBI works to establish a business model, IP over DTV via MPEG-2 packets should be a killer because of its higher bandwidth and data rate!

- In partnership with ISPs, broadcasters can transmit locally-cached unicast and multicast streams [19] over DTV spectrum rather than clogging more scarce wired TCP/IP infrastructure.

- Broadcast infrastructure can fully integrate into the Internet infrastructure and offer bandwidth of up to 150 Kbps that is particularly suited to UDP/IP multicast applications, and to push applications where a unidirectional stream can efficiently replace a large number of IP unicast streams.

- Programs developed for DTV/Internet applications can be scaled for a range of client devices. This may need multilayer programming.

- Distributed services could deliver part of the content via the wired Web and part of it via DTV.

5.7.2   Capitalizing on Business Opportunity

If Internet and PC clients form one of the endpoints of the emerging multimedia marketplace, perhaps broadcasting and TV clients can be seen as the other endpoint. Anything distributed via the Internet can be distributed via broadcast television—either via the VBI lines in the case of analog or via MPEG-2 packets in the DTV transport layer. For broadcasters to capitalize on this opportunity, they must (1) reconsider what their population of "client" devices is; (2) think of their signal in terms of data bandwidth and not TV channels; and (3) redefine their basic business model of traditional TV broadcasting—a single linear program with embedded advertising.

5.7.3   Redefining the Business Model

Broadcasters need to redefine their basic business model in a couple of ways. First, broadcasters are in the spectrum bandwidth business. One use of this resource is traditional television. Second, broadcasters need to redefine their retail distribution network from dedicated, single-purpose devices known as television receivers to anything capable of detecting, receiving, decoding, and processing encoded signals distributed via their spectrum emissions. This may include set-top boxes, PC-based devices, consumer electronic devices, cellular or PCS phones, and PDAs.

5.8   Conclusion

The examination of key drivers of ITV and related technologies show great promise. The world market is growing in terms of projected revenue in related technology and applications. The Internet, e-commerce, digital TV, cable, and satellite households are all projected to grow substantially in the next four years, according to several studies by survey and research companies. Digital broadcasting will be capable of providing ITV services in a big way. Client-server data flows are asymmetric. Client-side requests typically represent mouse clicks to select URLs or perhaps send e-mail. The higher bandwidth requirement is only toward the client. As such, DTV will be an ideal platform to deliver ITV. The ITV client devices must be user friendly, platform independent (in terms of API implementation), and cost-effective.

Once those requirements are met, ITV devices could support wide-ranging applications, leveraging both the powers of Internet client-server interactivity and DTV high bandwidth.

It might not make sense, economically or technically, to scale up the Internet infrastructure to handle relatively few peaks in events, particularly if there are local and regional spikes. Therefore, digital broadcasting will make up an important lane of ITV services. Collaborations and partnerships will provide win-win opportunities for ISPs, DTV operators, application and content developers, network providers, and, of course, customer care and billing agencies. As ITV providers envision their future growth, they have to ask themselves a few questions. First, what generic customer needs do they address with their current business model? Second, and more important to growth prospects, they have to address the issue of optimal utilization of assets that they currently ignore. Third, what business model is most suited for ITV services? Therein lie the seeds of a vast new market.

References

[1] *Communications Technology*, May 1994, p. 28.

[2] http://www.nua.ie/surveys/, May 2001.

[3] http://www.informationweek.com/story/IWK20010313S0006, May 2001.

[4] "Interactive TV and Home Control Set-Top Boxes: A Segmentation Analysis of Consumer Demand, Price Sensitivity, and Buying Intent," *TechTrends, Inc.,* November 2000.

[5] "Broadband E-Battle," *Deutsche Bank Alex Brown Research Report,* January 28, 2000.

[6] "AOLTV—Late to a Saturated Market," *Forrester Research,* June 19, 2000.

[7] "Portal Deals," *Jupiter Communications,* July 1999.

[8] Forrester's press release, July 23, 2000.

[9] "Global ITV Markets," *Jupiter Communications,* July 26, 2000.

[10] "Interactive Digital Television: Worldwide Market Forecasts," *Strategy Analytics* 2000.

[11] Forrester Research, Cambridge, MA, July 21, 2000.

[12] "ITVREPORT.COM: Interactive Television Industry Research," www.itvreport.com/news/research/current.htm, May 2001.

[13] "Television On-line Landscape," *Jupiter Communications,* January 4, 2000.

[14] "Smarter Television," *Forrester Research,* July 2000.

[15] *News Multichannel International: Fact Pack 2001,* Cahners, International Television Group.

[16] Nagaya, T., "Integrated Services Television: Digital Age TV with a Built-in Home Server," http://www.nhk.or.jp/, April 2001.

[17] *Multichannel News,* May 23, 1994, pp. 1.

[18] *CableWorld,* May 30, 1994, pp. 14.

[19] http://www.ipmulticast.com/, April 2001.

Selected Bibliography

"Consumer Interactive Services in Europe to 2002,"*Datamonitor Report,* December 1998.

http://www.datamonitor.com/dmhtml/dm/dmwtsnew.htm, April 2001.

http://cgi.zdnet.com/zdpoll/question.html?pollid=12167&action=a, May 2001.

"Interactive Digital Television Services in Europe and the U.S. in 2003," *Datamonitor Press Release* 10, May 1999.

"Mergers Within the Internet Industry Value Chain: Strategies for Success,"*Strategy Analytics—Press Release,* May 5, 1999: http://www.strategyanalytics.com/, May 2001.

"Set-Top Markets in Europe and the U.S.: Profit Opportunities in Digital Television," *Datamonitor Report,* December 1998.

Appendix 5.A

Table 5.A.1
Projected Investments for Building the Information Highway that Supports ITV and More*

Company	Investment (U.S.$ Billions)	Period (Years)	Incremental Investment (U.S.$ Billions)
Ameritech	33.0	15	4.4
Bell Atlantic	11.0	5	1.0
MCI	20.0	6	9.0
Pacific Telesis	16.0	7	2.0
SNET	4.4	15	0.4
Time Warner	5.0	6	3.2

* *From:* Wall Street Journal, May 18, 1994, pp. B1

Appendix 5.B

A number of companies claim that they are offering ITV, DTV, Internet TV, information appliances, or convergence technologies. Table 5.B.1 lists companies with their Web address.

Table 5.B.1
Web Sites of International Companies Dealing with Interactive Media

Company	Product services	Web site
Acer	Non-PC products	http://global.acer.com
ACTV	Enhanced TV, hyper TV, DTV, pro-gramming	http://www.actv.com
Alcatel	Broadband Internet, GSM, ADSL	http://www4.alcatel.com
Alex Informatique S.A.	High bandwidth, multimedia services, interactive learning	http://www.alex.com
allNetDevices	News and features for developers	http://www.allnetdevices.com
AOL	Web channels and ISP	http://www.aol.com
AOLTV	AOL features on TV	http://www.aoltv.com
Axcent	Media on demand	http://www.axcent.de
Boca Research	Modem	http://www.bocaresearch.com/index.html
BroadbandMagic.com	Broadband, converging technology	http://www.broadbandmagic.com
C-Cube Microsystems	MPEG video codec	http://www.c-cube.com
CableLabs	Set-top box and digital devices	http://www.opencable.com
CELERITY	Digital interactive services	http://www.celerity.com
Com One DomoTV	French site for ITV	http://www.domotv.com
Concurrent Computer Co.	VOD, real-time solutions	http://www.ccur.com
Continental Edison	Plasma TV, DVD	http://www.continentaledison.com
Coollogic	Embedded OS and Java	http://www.coollogic.com
Crosstainment AG (Surfstation)	Entertainment by TV/PC	http://www.crosstainment.com
Digital Video Arts	Digital broadband applications	http://www.dval.com
Eagle Wireless International	Wireless and DSL	http://www.eglw.com

Table 5.B.1 (continued)

Company	Product services	Web site
EnReach Technology, Inc. ("ippliance" solutions)	Integrates software, cable set-top box	http://www.enreach.com
Expanse Networks	TV advertising solution	http://www.expansenetworks.com
@MILYnet (la station multimédia)	Family TV	http://www.multimedia-nc.com
Fileants	ITV solutions	http://www.fileants.com
Flying Colors Interactive	Interactive content production	http://www.flyingcolorsinteractive.com
Focus Enhancements (TView I-Net box)	Enhancements, set-top	http://www.focusinfo.com
GCT Allwell Technology, Inc.	Set-top box, iDVD	http://www.gctglobal.com
GemStar International	Showview program	http://www.gemstar.co.uk
General Instrument	Broadband communications	http://www.gi.com
HTX Technology	Home automation, PC theater	http://www.hits.com
Humax	Korean software company	http://www.humax.co.kr
ICTV	Broadband TV	http://www.ictv.com
IDTV	iDTV	http://www.ihug.co.nz/idtv
IGS Technologies, Inc.	Broadband TV	http://www.igst.com
ImagicTV	Internet appliances, streaming media	http://www.tvia.com
ImaginOn	E-commerce, software	http://www.imaginon.com
Infomatec	Interactive services	http://www.infomatec.com
INO TVPC	TVPC and MP3	http://www.tvpc.com
Intellocity USA, Inc.	ITV provider	http://www.intellocity.com
21st Century Net, Inc.	Company for sale	http://www.2001cnet.com
ITV Report (Stanislav Levin)	Reports about ITV	http://www.itvreport.com
ITV Today (Tracy Swedlow)	Advisory and media co.	http://www.itvt.com
JCC USA I-Box	Provides Web site details	http://www.i-box.com
Kingston Interactive	Television iDTV over ADSL	http://www.kitv.co.uk

Table 5.B.1 (continued)

Company	Product services	Web site
Kingston Vision	Develops iDTV technology	http://www.kingston-vision.co.uk
Kingston Communication Group PLC.	ITV Service	http://www.kingston-comms.com
Liberate Technologies	ITV solutions	http://www.liberate.com
Loewe Xelos TVO	Virtual showroom	http://www.loewe.de
Lysis	Content management	http://www.lysis.com
M@ilTV	Allows e-mail on TV	http://www.mailtv.com
Matsushita Panasonic	Hardware company	http://www.matsushita.co.jp
Media Visions (Ken Freed)	iTV report	http://www.mediavisions.com/itv.html
Megabyte Inc. (MbTV)	Personal TV solutions	http://www.mbtv.com
Metabox AG	Digital set-top boxes	http://www.metabox.de
MeterNet	Church network	http://www.meternet.com
MiTAC	Wireless SOHO environment	http://www.mitac.com
Mitsubishi	Hardware company	http://www.mitsubishi.co.jp
Mixed Signals Technologies	iTV technology	http://www.mixedsignals.com
MoreCom's Internet TV	Part of Liberate	http://www.morecom.com
Motorola	Hardware, enhanced, handset	http://www.mot.com/home
MSU (SlipStream)	Web2U set-top box	http://www.msu.co.uk
MULTVmedia	Application provider	http://www.multvmedia.com
NagraVision	DTV and broadband Internet	http://www.nagra.com
NDS Ltd. (News Corp. subsidiary)	Cable DTV network	http://www.ndsworld.com/homepage.html
Netgem	DTV and Internet company	http://www.netgem.com
Orca Computers Ltd. (Israel)	Broadband, "RighTv"	http://www.orca.tv
Paradise Innovations (AiTV, WebEZ)	Multimedia Products	http://www.paradisemmp.com
PCTVnet ASA	IP Video Telephony	http://www.homepilot.com
Peach Networks	Taken over by Microsoft	http://www.microsoft.com/tv
PowerChannel FreePCTV	Channel provider	http://www.freepctv.net

Table 5.B.1 (continued)

Company	Product services	Web site
PowerTV	Developer of TV software	http://www.powertv.com
Rachis	System integrator of ITV software	http://www.rachis.com
Sejin America	Wireless keyboards for Sony, Hughes, Scientific Atlanta, EchoStar, Picturetel, etc.	http://www.sejin.com
Sigma Designs	Technology for digital video	http://www.sigmadesigns.com
SpotMagic	SpotTuner, live simulcast	http://www.spotmagic.com
Tcomnet	Internet appliance Internet TV	http://www.tcomnet.co.kr
TeleCruz	Internet TV solutions	http://www.telecruz.com
Telemann	Broadband solutions	http://www.telemann.com
Telelynx	CATV, LMDS, MMDS	http://www.telelynx.com.tw
TiVo	Live ITV with recording facility	http://www.tivo.com
TV Multicast	TV network	http://www.tvmulticast.com
TVPC	ITV service	http://www.tvpc.com
uniView	Digital set-top box	http://uniview.net
US Video Interactive	Internet video	http://www.usvo.com
Vestel USA	Information appliance	http://vestel-usa.com
VisionTech Ltd.	Broadband communication	http://www.broadcom.com
WavePhore WaveTop	Internet, digital solutions	http://www.wavephore.com
WebTV	WebTV of Microsoft	http://www.webtv.net
Wink	t-commerce via Wink	http://www.wink.com
WorldGate	Channel provider	http://www.wgate.com

6

Future of ITV

Within the next few years, 45 million broadband consumers will be looking forward to the innovative programming that merges broadcast video and Internet protocol to produce a participatory, nonlinear communications medium. Enhanced TV is in your future.

—The AFI-Intel ETV Workshop 2000's mission
statement (http://www.afionline.org/etv2000)

6.1 Introduction

Television has become part of our lives. It has helped shape the twentieth century, has become deeply entwined in commerce and politics, and continues to shape our personal views of the world. As of now, there have been several successful interactive products in the PC and game machine markets, and we are beginning to see the development of a significant market for the Internet on television. Costs of technology aside, early efforts to provide ITV services failed to spark great consumer interest. The reason for that failure was ITV's inability to deliver services consumers desired in a better way than they were already receiving them. Many of these early efforts shared the following common and critical mistakes:

- Concentrated on what the service provider wants to sell, rather than what the consumer wants to buy;

285

- Treated interactivity as a goal, rather than a method of achieving a goal;

- Treated consumers as part of a mass market rather than individuals.

The world in which ITV operates is changing rapidly. New technologies and new players have created a more complex and demanding marketplace for commercial television. The traditional values of public service broadcasting are under threat. Global, as well as domestic, pressures are forcing the ITV network, and the players that administer the services, to rethink their role. When we talk about the future of ITV, we need to ask ourselves a few questions. What exactly is entertainment? What uses do people have for TV? Can ITV's existing structure survive? How will it confront any sale of spectrum? Can ITV companies compete? Can they still claim to be public service broadcasters? What is commerce? Why do people want to purchase goods on TV without really touching and feeling them?

The nature of such questions spawns interest in the design, management, and production of media and smart products. This also depends on people, segments, groups, and communities and their behavior, social background, and purchasing capacity (Section 5.6). It is not uncommon for consumer products that are ultimately successful to initially be mispositioned. These mispositionings occur when developers leap ahead of what the technology can do or what they want to accomplish with the product, rather than start with a fundamental reexamination of consumer behavior and preferences. Consumption and usability are tied together in t-commerce, and people develop relations through and around media.

Innovation of new technologies always involves uncertainty and risk. This is especially true for radical new technologies where consumer or end user demand is unknown. Uncertainty about the future direction of ITV has not yet dampened the spirits of investors, who are apparently willing to forgo not only earnings but also revenue increases, as long as a company is positioning itself to be at the center of ITV. Innovation also makes firms reconsider their goals and skills, such as a broadcaster becoming a more general media provider. Innovation can bring great rewards, often unexpected, and under competitive pressure from other companies or alternative products and services, it may be more risky not to innovate.

Interactive television is a concept that requires innovations in many fields—technology, programming, services, markets, and organization—all against a background of unknown consumer demands. One of ITV development's strongest characteristics has been the formation of alliances, and massive takeover bids are now being mirrored on a smaller scale with the Internet (Section 5.5). This activity is a result of firms wanting to spread the risk of innovation (Section 5.4.1), and gather together the skills and resources of different sectors.

However, every firm has a different agenda and different expectations. This may be obvious between competitors, but ITV has illustrated many of the problems of trying to unite cultures and expectations of very different businesses. The success of those companies involved in this field will depend on their innovation process, the power and resources they have in alliances, their success in working with others, their reactions to external changes, and their vision of the future.

6.2 Why Interact?

Interactivity is not in itself desirable. It requires consumers to make an effort to receive content or a service. Why should consumers do that, unless the value received is substantially greater than the effort involved in getting it? If the interaction gives consumers results that more closely suit their personal needs or desires, they may be willing to make the effort. However, the interaction is bound to fail if it works on the one-size-fits-all philosophy. Each person is unique. ITV products have to serve the unique needs, interests, and tolerance for interaction of individuals within a common effort suitable for wide deployment. It must provide target users with an obvious benefit at an affordable price.

6.2.1 Interaction Modes

People have several modes of interaction, with different levels of motivation and tolerance for their extended effort (Figure 6.1):

- *Command mode.* The user gives an order and desires its immediate fulfillment (i.e., placing an order, making a purchase, or registering a complaint). Interactive product examples in command mode are all basic user interface command operations such as navigation, productivity applications, transactions, data retrieval, and device control. People in command mode focus on achieving their goal as quickly as possible. Their real desire is instant satisfaction, and their patience is very low. Efficiency, quick response, and a sharp focus on

S. No.	Mode	Attribute	Example	Characteristics
1.	Command	"Here is what I want; make it happen."	User interface command operation, navigation, data retrieval.	Low patience, quick result.
2.	Knowledge seeking	"And then what happened?"	Inquiring an expert, following a thread in Encyclopedia.	Patience and tolerance.
3.	Stimulation and self-expression	"WOW! Look at that."	Fun, creativity, toys and games, jokes, comedy, painting, drawing.	Absorption, no time limit.
4.	Community	"Participation, enjoyment."	Civic groups, churches.	Absorption, no time limit

Figure 6.1 Modes of user interaction.

moving quickly to the desired benefit are the keys to successful service of people in command mode.

- *Knowledge seeking mode.* The user seeks information (i.e., asking an expert, following a thread in an encyclopedia, participating in live question-and-answer sessions with follow-up). The process of knowledge seeking involves receiving a body of information, internally evaluating its fit into the person's existing knowledge, and then following up (with a question or seeking more input) to fill in missing areas or pursue related knowledge. As long as users feel that the information being received is interesting and following the desired thread to help them reach their goal, those in knowledge-seeking mode tend to be patient and tolerant.

- *Stimulation and self-expression mode.* The user needs fulfillment of creative desires (i.e., games, jokes and comedy, or "ear and eye candy" such as audio-visual special effects). Pursuing personal creative talents for enjoyment (painting and drawing, writing, music), creating a home page, being the center of attention when speaking to others, participating in forums, pursuing a hobby, undertaking a project, playing a musical instrument, creative cooking, and other such pursuits are examples of stimulation and self-expression activities. Users exhibit patience and absorption in this mode.

- *Community mode.* As social animals, we all have a fundamental need for relationships with each other, and much of our human-to-human interaction that does not fit into the modes above can be characterized as community. Community involves commonality—finding affinities with other people and creating a group of us that is set apart from the population at large. Some of these are formalized groups: churches, civic groups, political parties, clubs, and so on. Chat, group discussions, and religious discourses pertain to community mode. Much of the explosion of Internet use by consumers has been due to its ability to place people with common interests in touch with each other, without their having to leave the comfort and safety of their own homes. In community mode, people are very tolerant and patient, but only with those whom they consider within the same community. It is interesting to note that the actual behavior of people as they open a community-building conversation resembles those in knowledge-seeking mode: question, receive answer, evaluate, and formulate next question.

The modes described above apply generally to all humans, albeit they are expressed somewhat differently in adults and children. Young children will express their frustration to a failure in command mode, yet they can also be captivated by the novelty of a command mode device and be stimulated just by pressing the button to watch it work. Children tend to spend more time in knowledge seeking and stimulation and creativity than adults do. Although community is important to children, they are less us-them oriented with their peers than adults.

6.2.2 Human Behavior

Human values and institutions reflect and influence human behavior and inspirations, which are related to human biology, environment, culture, technology, and each person's individual experiences. New technologies affect human behavior and act as agents of social change. The development of writing, printing press, electronic communications, and others has influenced the behavior of mankind. ITV will also change people's behavior, thinking process, and living style. However, it will happen later. In the intervening period, developers and content providers will have to understand human behavior, their psychology, the need for intellectual and physical stimulation, and the constraints in which people operate. The technology, as well as the content, will have to be developed with these needs in mind. "Frightening" implications for viewer's behavior are likely to occur as ITV takes off, since consumers will have far more control over TV's programming. For example, ITV users in the United States are already using the technology to schedule programs and remove commercials (Section 2.9).

6.2.2.1   A Behavioral Model of Information Seeking

Just as animals evolve different methods of gathering and hunting food or prey in order to increase their intake of nutrition, humans also adopt different strategies of seeking information in order to increase their intake of knowledge. *Information foraging* refers to activities associated with assessing, seeking, and handling information sources, particularly in networked environments. Foraging for information on the Web and foraging for food share common features: Both resources tend to be unevenly distributed in the environment, uncertainty and risk characterize resource procurement, and all foragers are limited by time and opportunity costs as they choose to exploit one resource over another.

Research in organization science suggests that it might be helpful to distinguish between four modes of organizational scanning: undirected viewing, conditioned viewing, informal search, and formal search [1, 2]. In undirected viewing, the individual is exposed to information with no specific informational need in mind. Many and varied sources of information are used, and large amounts of information are screened. In conditioned viewing, the individual directs viewing to information about selected topics or to certain types of information. During informal search, the individual actively looks for information to deepen the knowledge and understanding of a specific issue. During formal search, the individual makes a deliberate or planned effort to obtain specific information or information about a specific issue. Search is formal because it is structured according to some preestablished procedure or methodology. Table 6.1 below identifies the four main modes of information seeking described above. For each mode, the table indicates which information seeking activities or moves are likely to dominate, as suggested by theory.

Table 6.1
Information Seeking Modes

Scanning Modes	Information Need	Information Use	Amount of Targeted Effort	Number of Sources	Tactics
Undirected viewing	General areas of interest; specific need to be revealed	Serendipitous Discovery, "Browsing"	Minimal	Many	Scan broadly a diversity of sources, taking advantage of what is "easily accessible," "Visioning"
Conditioned viewing	Able to recognize topics of interest	Increased knowledge about topics of interest, "Learning"	Low	Few	Browse in preselected sources on prespecified topics of interest, "Discriminating"
Informal search	Able to formulate queries	Increased knowledge on area within narrow boundaries, "Selecting"	Medium	Few	Search is focused on area of topic
Formal search	Able to specify targets	Formal use of information for decision, policy-making "Retrieving"	High	Many	Systematic gathering of information about an entity following some method or procedure, "Optimizing"

6.2.2.2 Ellis' Model of Information Seeking Behaviors

Ellis (1989), Ellis et al. (1993), and Ellis and Haugan (1997) propose and elaborate on a general model of information-seeking behaviors based on studies of information-seeking patterns of social scientists, research physicists, engineers, and research scientists [3–5]. The model describes six categories of information-seeking activities:

- *Starting:* Forming the initial search for information, such as identifying sources of interest, which are likely to point to, suggest, or recommend additional sources or references;

- *Chaining:* Following up on new leads from an initial source (can be backward or forward);

- *Browsing:* Searching in a semidirected fashion in areas of potential search;

- *Differentiating:* Filtering and selecting from among the sources scanned by noticing differences between the nature and quality of the information offered;

- *Monitoring:* Keeping abreast of developments in an area by regularly following particular sources;

- *Extracting:* Systematically working through a particular source or sources in order to identify material of interest.

Table 6.2 relates the four main modes of information seeking on the Web (Section 6.2.2.1) with Ellis' model of information-seeking behavior. For each mode, the table indicates which information-seeking activities or moves are likely to dominate.

6.2.2.3   Basic Human Needs

All humans have some basic needs (i.e., hunger, shopping for essentials, or entertainment). Fulfilling these needs helps people get over their natural fear of intimidating new technology. In the initial stages, and perhaps afterwards, methods to fulfill these basic needs are more likely to be popular with the masses. "Because people want their news and sports for free. One of the keys to our success on the Internet is that we've never tried to follow an advertisement-supported business model," says Ken Boenish of New

Table 6.2
Modes of Seeking Information on the Web

Starting	Chaining	Browsing	Differentiating	Monitoring	Extracting
Undirected Viewing	X	X	—	—	—
Conditioned Viewing	—	—	X	X	X
Informal Search	—	—	X	X	X
Formal Search	—	—	—	X	X

Frontier Media [6]. The site is pay-channel, but popular because it fulfills human needs.

When developing ITV programs, providers should keep in mind the need to make them more popular. TV chats, TV videophones, and instant e-mail/voicemail that helps people meet on-line while watching live video are likely to be popular while also fulfilling people's emotional needs. The ability to reach out and touch someone via a TV will persuade many to embrace ITV.

6.2.2.4   Framework for Information Intake

People need information and entertainment for intellectual, as well as physical, stimulation. There is a segment of people who need knowledge, education, and information to quench their curiosity and achieve mental stimuli. There is another segment that needs information and entertainment for physical ecstasy. External elements, in the form of the following constraints, also affect our intake of information and entertainment:

- *Social constraint.* Since the human being is a social animal, information selection behavior is restricted by social conditions. Social constraints limit the choice of viewing a particular type of program in a home or office. Many offices have placed restrictions on computer use, including e-mails.

- *Time constraint.* This limits the viewer's patience when searching a program or even watching a program at a specified time.

- *Economic constraint.* This determines the service's affordability, the end terminal's purchasing power, and even the amount of teleshopping.

- *Constraint of intellectual interest.* This limits the type of content offered on the ITV. A few people may like light entertainment, while others may like a mentally stimulating quiz program. It is interesting to know that studies by Miller in 1966 revealed that the number of information chunks that one can simultaneously distinguish and memorize is limited to about seven. All these segments will need targeting with personalized programs.

Figure 6.2 shows a framework of information intake under various constraints. It can be seen that information uses are subject to internal and external constraints. There are limits to how people can allot their time, money, and interest to information-related behavior, and that the dimension in this regard has physical limitations because a person is a physical being.

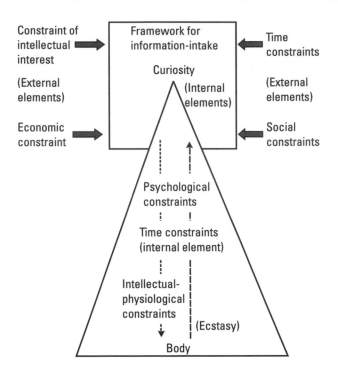

Figure 6.2 A framework of information intake under various constraints.

Apart from the external constraints mentioned above, people have internal constraints based on their psychology and physiology. Time as an internal constraint is based on people's behavior. People may have the time but not the patience or interest for a particular type of program. Under all these constraints, people still want to fulfill their individual needs of mental curiosity and ecstasy (body). All these constraints and needs are different for individuals. ITV content needs to especially target audiences with varied needs and constraints. ITV addresses many of these issues by providing personalized services at user-specified times, but it also has to address those issues related to economic constraints.

6.3 Contextual Usability

The life cycle of ITV product development and its marketing is shown in Figure 6.3. It can be seen that during Phase 1, designers and developers carry the product through constant iterations. In Phase 2, content developers

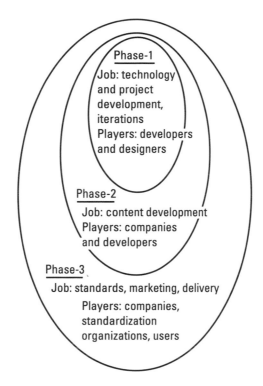

Figure 6.3 Phases of product life cycle.

provide appealing and innovative content. Thereafter, the focus shifts to standardization, marketing, and product delivery to users. In the case of ITV, the technology is in place, albeit it will continue to develop and throw new challenges for all players (Section 2.8) in the second and third phases. However, with the technology available, content developers and network providers can administer very promising ITV services.

Standards organizations (such as the ITU and DAVIC), central institutions in the mediation of technical change, are key points of contact between otherwise competing firms (Section 3.7). Universities and independent research organizations are important centers of experimentation and public information, both in technology and in creating content. In Europe, the current emphasis is on MHP Version 1.0 (Section 3.7.1.7). However, work is progressing on MHP Version 1.1, which will support the XML and HTML 4.0 protocols favored by ATVEF (Section 3.7.5.4). Down the road, MPEG-4 will support advanced interactivity [7] (Sections 1.4.2.3 and 3.3.2).

Standardization is very important in the growth of ITV. To date, most providers are using proprietary software. For example, in France, the two major providers of ITV services are digital DTH platforms Television Par Satellite (TPS) and Canal Satellite, and both use proprietary software. TPS is deploying OpenTV middleware and Canal Satellite is deploying its own MediaHighway system.

Governments, even in an increasingly deregulated industry, still maintain a central role (Sections 3.10 and 3.11). Today, however, governments are moving to consolidate policies toward new TV services with those of other information and communications technologies (ICT), as they realize that mass-market entertainment and advertising are key factors in bringing multimedia technologies and services to the general public. The deregulation of the telecommunications industry has been a key factor in stimulating ITV. Now, tweaking regulations and legislation makes governments the center for huge lobbying and can change the economic balance in favor of one technology or another, or one company or industry committed in ITV against another. In addition to the regulation issues, the idea of the "information society" or the "information infrastructure" aimed at national competitiveness and bringing technical solutions to social problems has been an important backdrop for awareness of new technology, investment, and regulation priority. The TV has great social impact (Section 1.7). Interactive television, with its connotations of entertainment, commercialism, and lack of economic instrumentality, has to be an important part of that future vision.

The purest form of ITV will integrate brand marketing with sales and impulse purchase. The key is synergy, and the key to synergy between the Web and ITV is intelligent marketing. ITV can serve up communications intelligently, with perhaps different content communication on the Internet. Both the Web and ITV are just digital vehicles. Together, they can deliver the same content with much greater impact. But the TV is the best way to "get people in" [8].

Creative Challenges

Where content is king, content creation is the jewel in the crown.
    —Charles Allen, executive chairman of Granada Media,
        in *Management Today* (December 2000)

ITV throws great challenges to content providers, developers, and marketing teams. In Section 6.2, we examined the information needs based on human behavior. The content has to be innovative and appealing to the target segment (Section 4.2). There are difficulties surrounding good and appropriate ITV concepts, design, and navigation. There are also challenges facing consultants, producers, designers, and developers in creating good-looking, intuitive services that fulfill their objectives, whether for business or purely for entertainment. For ITV to be successful, the aspects discussed above will have to be kept in mind when creating compelling content. Commercials will also need special attention. The era of 30- or 60-second advertisements will be over. There may be hot contents embedded in the main program (Section 4.2.2). ITV also raises a number of issues in regard to privacy, culture, impact on users, and many more.

Content

In 1995, *Mr. Payback,* a low-budget, 20-minute movie, became the first interactive film in history. The theatre was wired and users were provided a voting device where they could enter one of three options to decide what would happen next in the film. A special console was provided to capture user responses. Approximately once every 60 seconds, moviegoers could select the option. In total, nearly two hours of footage was shot and a viewer could watch the movie 25 to 30 times before it became repetitive [9].

ITV has the potential to personalize television to each individual (Section 2.5). The most significant driving forces in content creation are the perceived wishes, hopes, and expectations of the audience. This means collecting data about users habits and behavior (Section 1.11). The set-top units have the ability to register all the interactivities a TV viewer registers. Network operators have the potential to sell information about how often and how loyal a viewing consumer is to programmers, marketers, and advertisers. As a method of determining who is watching and for how long (as well as viewer attentiveness) such audience information offers great promise to content providers and advertisers. The obvious attraction of ITV is the fact that viewers will react to certain advertisements, network service providers can aggregate this data and then contribute to our understanding of the changing patterns of consumer responsiveness and buying behavior. Many have picked up on this point already. Yet few have identified what is possibly the more golden opportunity—namely, the ability to track viewing patterns.

Programs may be canceled not because of low ratings but rather due to lack of viewer interactivity. The effectiveness of the TV commercial will be consistently tested. The 30-second commercial may even become a thing of the past, since users can skip commercials with a smart TV (Section 2.5). New ways, such as embedding commercials in the main program, are being tried

(Section 2.4.8.4). Cost-per-response may replace present modes of TV program ratings as the basis of negotiation. Another ITV potential is programming specifically designed for interactivity. Award shows in which the audience selects the winner may also become popular (Section 4.3.3.1).

Clients are looking to agencies to provide a multitude of new disciplines, while at the same time having to combine traditional with new media agency experience. The ability to harness and understand ITV and deliver projects integrated with a range of new technology, so that the program or company's brand is imprinted on the conscience of the viewer, is very important. Agencies and production companies need to find innovative, creative ways to grasp ITV and new media in order to communicate an effective message to customers. What will be the killer application for ITV? What are the driving forces—the killer applications—that have so far eluded ITV? Figure 6.4 shows the classification of a few programs with reference to the human interactivity mode discussed in Section 6.2.1.

Examples of command mode ITV programs are on-line transactions (such as ordering, banking, stocks, and travel). Other uses in this mode are data retrieval (looking up a particular company or piece of information, tracking stocks, delivery of news summaries, frequently asked questions, etc.), customer service, and software downloads. A slow starter, command mode ITV programs are continuing to grow as more and more products and services come on-line, and as people become more accustomed to thinking of on-line as a viable alternative.

The knowledge-seeking mode mainly involves browsing and exploring an area of interest or looking for new and interesting information. This is the most common use of the Internet. With more than 1 million sites on the Web, there is enough content that virtually any user can find new and useful information suited to his or her unique interests. However, with the Internet being a big maze, finding the desired information is always a laborious job. ITV has to address this issue. Interactive news is one example of ITV in this mode.

Stimulation, creativity, and self-expression are the most sought-after programs in ITV (Section 2.4.8.5). Generally, the games played on-line do not compete with video and computer on a stand-alone basis, but when the game can be played with others in multiplayer mode, on-line offers opportunities that stand-alone games cannot provide. Gambling is a wonderful fit. Through ITV, viewers can watch sporting events on the main screen and order up bets on a picture-in-picture screen. It can all be done through menus accessible by clicking the remote control. Developments in multiplayer gaming over two-

Program	Classification	Remark
Video-on-demand	Command mode	—
Interactive channel guide	Command mode	—
Games and puzzle	Stimulation mode	—
Movies with steering wheel	Stimulation mode	Users to change outcome of story
Home shopping	Knowledge searching, Command mode	—
Interactive news and documentation	Knowledge-seeking mode	—
Home banking	Command mode	—

Figure 6.4 Classification of some programs with reference to human interactivity modes.

way communications networks will lead to greater interest and increased subscription rates for ITV gaming. But until broadband Internet access is widely adopted, most server-side gaming services will consist of low-bandwidth, turn-based games, such as chess, cards, game shows, and puzzles. These types of games appeal to the mass market, providing service operators with numerous revenue-generating opportunities.

There are thousands of on-line opportunities for users to express their own opinions and contribute to the growing content base. A person's self-expression that is published on-line is then available to the world (or to everyone within the service).

Commercials and Advertisements

New, creative concepts can be applied to advertisements currently based on the 30-to-60-second hit-and-run philosophy. The ad can sit behind the scenes, the viewer can wander off with the ad, or the viewer can view several content sources at once. Data can be gathered regarding the brand of car that a viewer likes, and then car ads can be delivered that are tailored specifically for that person. Advertisers can also monitor which ad viewers are particularly interested in and react accordingly. But the creative concept will be what gets people's attention in the first place.

The possibilities for brand marketing are particularly fruitful. At present, brand marketing is one of the most nebulous concepts for TV advertising, requiring long, intensive media schedules hooked closely with retail spots. But ITV will make brand marketing more effortless. The closest example is a Pantene shampoo ITV ad now running in the United Kingdom, which goes through nine questions on hair type with the viewer and provides specifically tailored advice and information. The whole experience takes 10 minutes. It is getting a 45% response rate, which is phenomenal. How many traditional advertisers get to spend 10 minutes with a viewer and present their brand as a quality authority in the category, as well as promote specific products? ITV will allow intelligent dialog, which builds brand in a very startling way. ITV gives advertisers more control of what viewers are watching.

Data Mining and Privacy Issues

The most important data, which the ITV broadcaster can collect and use to build new target segments, is information on what each household watches. Web site managers have been using "cookies" to do this on a widespread basis for some time. This data is factual, in that it is a real record of the household's interests as reflected by their viewing patterns. It provides both intuitive indicators and the foundations of a deeper psychographic picture of the viewer. It is also a wealth for targeting viewers. Regular watchers of gardening programs are good targets for a seed catalog operation. The combinations of programs that people watch are good indicators of their aspirations and attitudes. Psychographic system providers often describe their different social value groups in terms of the TV programs they favor. The descriptive power of this data is, therefore, of quite staggering significance to the marketing professional.

Interactive household viewing data is geographically referenced. It refers to a specific television, in a specific household, at a specific address.

Viewing data can be combined with lifestyle, census, and transactional data. Many network operators have collected data mines of interactivity data from viewers. This raises a number of privacy issues. Viewers do not always want their information collected or used to target them with advertising or other messages. They should be able to control whether this is done or not. At present, most ITV suppliers are reluctant to address the privacy issue. In the future, serious thoughts will be needed to address several privacy issues.

6.4 In the Research Lab

There are several R&D efforts to explore new ITV services. These are based on a combination of DAB (Section 1.6.10.1), DMB (Section 1.6.10.3), and DVB (Section 1.6.1) with fixed and mobile telephony (Section 3.2). Efforts are also being made to use Ka band for the return channel for ITV applications, whereas the forward channel uses DBS (Sections 1.6.6 and 3.8).

6.4.1 DAB for Traffic, Travel, and Tourism Information

DAB is a wide-bandwidth broadcast medium and presents opportunities to develop new interactive traffic system (ITS) applications. Table 6.3 provides the flexibility of DAB in comparison to other broadcast techniques.

Table 6.3
Flexibility of DAB Technique

Technology	Data Rates	Mobile Use	Uni-/Bidirectional	Projected Life Span (years)
DAB/DMB	1.7 Mbps	yes	unidirectional	20
GSM	9.6 Kbps	yes	bidirectional	10
UMTS	384–2000 Kbps	mobile-stationary	bidirectional	15
Satellite-based service	2.4 Kbps (initially)	portable	bidirectional	10
FM	1187.5 Bps	yes	unidirectional	+10
DARC	10 Kbps	yes	unidirectional	+10
DVB-T	5–31 Mbps	mobile-stationary	unidirectional	20

From the above table, it is evident that the advantages of DAB, such as excellent mobile reception and data services, promise mass-market potential. The high data rates allow for a range of features, such as:

- Flexibility of bandwidth use;

- Flexibility of receiver design, from the most complex to the simplest (more intelligence in the infrastructure and less in the receiver means cheaper receivers);

- Dynamic navigation when linking an in-vehicle navigation system to live traffic information;

- On-line transmission of digital maps and map updates (high data rates);

- Very precise positioning [transmission of global positioning system (GPS) location signals];

- Updates of location tables;

- Parking reservation and pretrip information when combined with a return channel and communications system.

Combined with global systems for mobile communications (GSM) (Section 3.2) or GPS, DAB can support a range of ITS services for pretrip information, route guidance, and dynamic navigation. These combined services should make traffic and travel information, together with tourist and weather information, available at the home, in the office, in a vehicle, and at kiosks. This would enable travelers to plan and compare trips that use a combination of transport modes. On-trip services could range from traffic information, which can be filtered for the trip, to public transport information. The value-added services could be reserved and paid for.

The European potential market forecast for advanced traveler information services, such as in-car navigation and traffic information, is 26 billion ECU per annum.

6.4.2 Data Delivery Services

DAB's data downloading capabilities open up the potential to deliver many sorts of data to customers, such as digital map updates, yellow pages, software upgrades, broadcast Web sites, the Internet, and personal files. There is also potential for commercial vehicle document clearances, fleet management, and car pooling services.

6.4.3 IDMB

In conjunction with the GSM+DMB hybrid system, Bosch GmbH is developing a hybrid system combining DMB and GSM cellular technology. The transmission technology itself and the terminal devices, such as the TravelPilot navigation system, are now becoming practical for use in automobiles. The new hybrid system will also integrate the next generations of the cellular phone system—the General Packet Radio system (GPRS), with a bit rate of 115 Kbps, and the Universal Mobile Telecommunications system (UMTS), with a bit rate of 2 Mbps. These values apply to a stationary vehicle. With the vehicle in motion, they are substantially lower, so a DBM module is desirable. Another factor favoring the use of DMB for the downlink is that broadcasting services cost less than cellular phone communications and are likely to continue to be less expensive for the foreseeable future [10].

6.4.4 Two-Way Broadband DBS

Ka-band (20–30 GHz) systems can support full broadband Internet access, full-function VOD, enhanced and individualized programs, t-commerce, and even multiplayer games. Two-way Ka-band positions DBS (Section 1.6.7) in the same league as digital cable, which holds onto a thin lead in signal response time. Every user can receive the multimedia data via satellite in Ka band and interact using a very low-power transmitter available in his or her own home.

iSKY [11] in Denver plans to launch two-way Ka-band satellite services by 2002. Starting with broadband Internet before moving into interactive TV, iSKY is planning to deploy three satellites covering North, Central, and South America. Global reach is contemplated. iSKY has an exclusive license for commercial Ka-band capacity on the TeleSat ANIK F2 satellite at 111.1 degrees West longitude. Loral Space Systems is constructing the iSKY's first Ka-band satellite. The satellite's spotbeam will interact with clusters of home transceiver dishes within a 400-mile footprint per beam. iSKY's rectangular 26-inch dishes will be able to receive signals from multiple satellites through a DOCSIS modem in the set-top box.

iSKY feels that the U.S. market for Ka-band Internet services is between 25–30 million customers. DIRECTV and Dish Network are also expected to launch Ka-band bidirectional satellites for ITV and t-commerce.

6.4.5 Emerging Services

There are already vast global networks linking various business, academic, and domestic communities. Speech recognition technology is also advancing, enabling even more sophisticated, yet easy-to-use, interfaces. As the home gets wired up for full high-speed two-way communications, many more applications will become feasible, such as video conferencing, collaborative working, telecommuting (working from home), video-mail, video-telephony, telepresence (virtually being there when you are not), telemedicine (remote diagnosis and specialist consulting), and distance learning [12]. The ITV terminal may become an interface for all these services.

6.5 Gazing into the Crystal Ball

What will a world be like in which the power of ITV is available to every home? Schoolchildren will be taught to write multimedia essays using clips from current news programs and material researched from on-line libraries and video servers around the world. Customers' comments will be filmed and transmitted direct to the design team. Fans will follow the daily schedules of their particular obsessions, whether that is pop stars or batsmen, receiving broadcast messages and other information at carefully planned intervals. Interactive television will transform education. The rate of knowledge advance will accelerate dramatically. Passive receiving devices (TVs and radios) will gradually give way to active digital devices. Some will be based on PCs; some will be built by adding a digital box to a conventional TV set; some will consist of integrated intelligent TVs. On-line marketing will be a growth industry, with companies offering to put together buyers and sellers [13]. Let us try to look into the crystal ball of ITV:

- Public service broadcasting will continue to play a key role in the digital future, potentially even more important than its current role. However, the way public service broadcasting is regulated and delivered by broadcasters will have to change to reflect the new conditions in which they operate.

- TV companies will emphasize the Internet's open network protocol, TCP/IP, instead of its display mechanism, HTML.

- Currently, the airwaves can only handle a small amount of data. Digital TV will offer higher packaging capacity. Broadcasters will carve up the HDTV spectrum to stretch transmission to about 20 Mbps. They will use it to deliver data, as well as high-resolution images, to consumers.

- Broadcasters will carve up the HDTV spectrum and deliver better images over existing TVs, using the rest of the spectrum for more channels and data. HDTV spectrum will become the t-commerce network backbone.

- DTH service will provide broadband interactivity. With the recent adoption of specifications for interactive channel satellite distribution, DVB-RCS by European Telecommunications Standards Institute (ETSI), the satellite-based delivery of ITV services will get a boost. (The specification ETSI EN 301 790 can be downloaded from the Web site, http://www.etsi.org/.) The Societe Europeenne des Satellites (SES) is already implementing "the world's first interactive satellite network based on DVB-RCS, the Astra Broadband Interactive (BBI) system, which uses Ka-band payload on the Astra IH satellite. BBI enables the user to benefit from a high-peak data rate on the forward and return link, integration with DTH broadcast, independent from terrestrial infrastructure, and the inherent multicast support from the satellite" [14].

- The mobile phone will play an important role in interactivity. Some of the lab experiments were described in Sections 1.6.7 and 6.4. With Java-enabled handsets, third-party application developers will have access to the mobile phone platform. Mobile subscribers will be able to download games and applications on to their handsets. These will run locally using the handset's memory and processing power [15]. By connecting a mobile to various sensors, the phone can be turned into a device that could measure heart rate, send data to someone, and seek advice. The integration of these mobile devices with ITV content will generate new revenue streams.

- Advanced radio frequency (RF) handheld devices will become available. These devices will be able to control household entertainment units via a touch-screen pad that will also serve as a monitor. TV viewers will be able to download interactive contents directly from a digital set-top box. Universal Electronics, Inc. (UEI) [16] demonstrated such a unit at CCTA Western Show 2001, at the Los Angeles Convention Center. According to the manufacturer, the device allows users to control up to 15 audio-visual devices by using different screens that can be personalized in order to access any home entertainment system [17].

- As an open medium for all users and suppliers of information, the Internet will make a great range of services available to the mobile customer. Hotels or restaurants, for instance, will be able to encode their local coordinates on their Web page, which—when accessed by the navigation system—will automatically guide the driver to that destination. The IP Wireless Application Protocol (WAP), used on cellular phones, should therefore also be standardized for location-specific data. New vehicle functions such as remote diagnostics and remote service are now entering the realms of motoring reality. In the event of a breakdown, the motor vehicle is able to contact a service center automatically via the Internet. In turn, the service center diagnoses the fault and then directs the vehicle to the nearest repair shop or notifies the breakdown service.

- While click-through to Web destinations via current set-top box technology is now available, the Web as an ITV destination is a short-term model. The real future will lay in ITV-specific content and delivery channels, where the Web and Web marketers will provide database management, and perhaps even content provision, roles.

- Instead of TV programs telling customers to visit a particular Web site, within five years the majority of TV channels will have embedded Internet content, and the TV experience will be highly personalized.

- One possibility is that e-commerce will become a B2B medium, while t-commerce will be a B2C medium.

- On-line marketers already skilled at forms of interactivity will enter the fray. They will provide free content but also sell products and service via a Web site, becoming in effect both a supplier and delivery channel or content provider.

- ITV companies will continue as the main commercial providers of public service broadcasting, before and after digital switchover, but with less prescriptive, detailed regulation.

- TV is poised to be the next major interactive advertising and t-commerce portal. T-commerce will succeed where ITV failed. Lawrence Marcus, author of the Deutsche Bank report and the man responsible for coining the phrase "t-commerce," says the introduction of interactive media and t-commerce will be worth $300 billion over the next five years. "Five years from now, t-commerce revenue is going to be greater than e-commerce revenue," says Mr. Marcus. Further, home shopping through TV shopping will be worth $65 billion by 2002. Interactive advertising will be worth $201 billion, music and movies on demand will be worth $29 billion, and games will make up the rest.

- ITV networks were built on proprietary network protocols. There was lack of content, few third-party providers, and the boxes were prohibitively expensive. The ITV will provide the t-commerce infrastructure and content, and the boxes will cost less than $300. Furthermore, the old ITV was focused on VOD—getting the program you want when you want it. The new ITV will focus, at least for the time being, on commerce

- Advertising within actual programming will become more sought after. That means companies will pay to have a character in a movie drinking a brand-name soda or wearing brand-name pants [18].

- Many homes will have the TV as a nerve center or as part of a networked home, controlling devices and appliances throughout the home. There may also be a possible demand for TVs networked in the home like in a TV LAN package, with TVs throughout the home linked to a central set-top or computer in the home. The future could be more of a TV-oriented world than a PC-centric world. However, there will always be PCs, and many of them will be entertainment PCs or may even become servers on possible TV LANs in networked homes. In the not-too-distant future, the television could become the multimedia hub of the household, with connectivity to multiple household devices via the IEEE 1394 digital interface.

- The set-top box has to go away. All of the different elements inside the set-top box will migrate into other devices in the home or be hidden. The next-generation enhanced TV devices will be standards-based and include as de facto components HTML-compliant browsers, hard disks and caching, sophisticated graphics, and more memory to be able to run middleware like Sun Microsystems's Java Virtual Machine.

6.6 Conclusion

ITV promises great potential. The market will grow over time, with viable segments using telephone, cable TV, satellite, digital broadcast, and mobile and cellular connections. Cable will be a driving force but may not totally dominate the market. Other means of connectivity, such as a combination of DAB and GSM, will provide a few killer applications. Consumers' habits will also play a major roll. Consumers may not want to deviate too much from their existing TV viewing habits. They may not want to use their TV sets as computers, and they may not necessarily want full PC-TV sets or computers that happen to have TVs as big monitors. A large segment of consumers may want to use uncomplicated, basic Web functions on their TV sets, and they may just want to watch TV, whether that watching consists of regular TV shows or of interactive information content that they can easily surf or access without too much hassle. There will also be consumers who desire to have TV sets with all the complexities and associated potentials. This will likely happen more so with the rapid increase in computer literacy around the globe and the ease with which consumers handle complicated gadgets in everyday life. In the future, consumers will be able to buy a lot of products via TV interactivity. This may also be more likely with the younger generation, who will have greater monetary affluence as the world becomes more prosperous.

Although both the Internet and ITV will be under construction for many years to come, this will not prevent their active application in many areas. Digital highways can—and will—be used even while they are being built, and construction will continue even when the network is fully deployed. ITV will play an important role in learning and education. One school in the United Kingdom, Netherhall School in Cambridge, is already using and creating ITV as an integral part of its learning program.

The future is exciting, though it is not for the fainthearted. This is a brave, new world, and we have two options: cling to the past, in a world that is being torn apart and reassembled, piece by interactive piece, or embrace the future. We can place our heads in the sand and hope for the best. Or, we can accept the challenge of learning what it takes to survive and prosper by new rules. The rewards, for those who take this step, are there for the taking.

References

[1] Aguilar, F. J., *Scanning the Business Environment*, New York: Macmillan Co., 1967.

[2] Aguilar, F. J., *General Managers in Action*, New York: Oxford University Press, 1988.

[3] Ellis, D. and M. Haugan, "Modeling the Information Seeking Patterns of Engineers and Research Scientists in an Industrial Environment," *Journal of Documentation 53*, No. 4, 1997, pp. 384–403.

[4] Ellis, D., D. Cox, and K. Hall, "A Comparison of the Information Seeking Patterns of Researchers in the Physical and Social Sciences," *Journal of Documentation 49*, No. 4, 1993, pp. 356–369.

[5] Ellis, D., "A Behavioral Model for Information Retrieval System Design," *Journal of Information Science 15*, No. 4/5, 1989, pp. 237–247.

[6] Cosper, A. C., "Leveraging Content," *Broadband Satellite*, April 2001, p. 6.

[7] Bayet, E., "France Telecom Unveils DVB-MHP Test Platform," *Multichannel News International*, April 2001, p. 32.

[8] http://australia.internet.com/r/article/jsp/sid/432121, April 2001.

[9] http://www. colossus.net, April 2001.

[10] http://www.bosch.de/bri/bri_e/, April 2001.

[11] http://www.isky.net/, March 2001.

[12] http://www.strl.nhk/, March 2001.

[13] http://www.mediation.co.uk/, April 2001.

[14] Television Omnibus, "Return Channel Specifications Adopted," *World Broadcast Engineering*, Vol. 24, No. 3, March 2001, p. 12.

[15] Chan, T., "Brewing Up an Open Platform for Mobile Phones," *Wirelessasia*, January/February 2001, p. 40.

[16] http://www.uei.com/, April 2001.

[17] http://wireless.newsfactor.com/perl/story/5549.html, April 2001.

[18] Jakel, P., "Bonefire," *Broadband Satellite*, February 2001, pp. 20–24.

Selected Bibliography

"Broadcast Satellite Services," http://www.isp-sat.com, April 2001.

Chankhunthod, A., et al., "A Hierarchical Internet Object Cache," *Proc. 1996 USENIX Technical Conference*, San Diego, CA, January 1996.

http://moneycentral.msn.com/articles/invest/sectors/5053.asp, April 2001.

http://www.chyron.com, May 2001.

http://www.office.com/global/0,2724,54-22339,FF.html, December 2001.

http://www.the-arc-group.com/press/interactive_tv.htm, April 2001.

http://www.upside.com/texis/mvm/news/news?id=394fe8e80, April 2001.

http://www.w3.org/Architecture/1998/06/Workshop/paper08/, December 2001.

http://www.wbeonline.com, April 2001.

Nonnenmacher, J., and E. W. Biersack, "Scalable Optimal Multicast Feedback," *Institute EURECOM*, BP 193, 06904 Sophia Antipolis, France, July 1997.

Nonnenmacher, J., E. W. Biersack, and D. Towsley, "Parity-Based Loss Recovery for Reliable Multicast Transmission," *SIGCOMM '97*, Cannes, France, September 1997, pp. 289–300.

Rodriguez, P., E. W. Biersack, and K. W. Ross, "Improving the Latency in the Web: Caching or Multicast?," *Third International WWW Caching Workshop*, June 1998.

SkyCache, http://www.skycache.com, April 2001.

Television Omnibus, "Moving Broadcast Quality to Interactive TV," *World Broadcast Engineering*, Vol. 24, No. 4, April 2001, pp. 14–15.

www.cnn.com/2000/TECH/computing/12/11/microsoft.interactive.tv.idg/, April 2001.

www.digitrends.net/marketing/13642_10927.html, March 2001.

www.interesting-people.org/archive/1326.html, March 2001.

www.microsoft.com/msft/speech/DevaanING00.htm, March 2001.

www.opentv.com/industry/tvexpert/n_headlines.html, March 2001.

www.zdnet.com/zdnn/stories/news/0,4586,2440672-1,00.html, April 2001.

Appendix 6.A

ITV and Related Technology Development Timeline

Development During 1998–2000

Year 1998

Households with DTV who like to receive DTV:

- Via DTH   95%

- Via cable   5%

Worldwide:

- Digital set-top boxes (production)   13.3 million
- Internet game devices   848,000
- Handheld devices   1.6 million

Households with:

- 3+ televisions   54%
- PC   48%
- 2+ PC   12%
- Internet access   21%
- E-mail users   43%
- Planning PC-TV   4%
- Sales for overall consumer electronics $75 billion
- Global market for information appliances   5.9 million

Receiving Internet-based information for:

- News   37%
- Movies   20%
- Dramas   10%
- Sales of set-top boxes   11.4 million

Year 1999

- TV households in United States   110 million
- Adults on-line   79.4+ million
- People planning to go on-line   18.8 million

Worldwide:

- Digital set-top boxes (production)   19 million
- Internet game consoles   5.9 million
- Handheld devices   3.7 million

Worldwide households:

- Watching TV and working simultaneously   18 million
- Having a PC and TV in the same room   5 million
- Networked homes   100,000
- Sales for overall consumer electronics $80 billion
- Devices, including set-top boxes, handheld computers   11 million
- Revenues from the market $2.4 billion
- Digital/personal video system   300,000

Year 2000

- Adults on-line   100 million

Households:

- Connected to Internet   49%
- With PC   42.7 million
- With modem   40 million
- Using Internet   36.5 million
- Using Internet appliances   12.2 million
- Worldwide PC sales to overtake TV sales
- Home network market $1 billion
- ITV audience growth   180%
- Homes with access to TV shopping, banking, e-mail   16.9 million
- Homes with DTV   56 million
- Digital set-top box production   25.1 million
- Multimedia industry $10.5 billion

Development During 2001–2005

Year 2001

Worldwide:

- Cable and set-top boxes   104 million
- Game consoles   105 million
- Entertainment PCs   150 million
- DVD players   31 million
- Internet TV sets   1 million
- DTV sets   1 million
- Market for low-cost network appliances and PC product $5 billion
- ITV   12 million

Year 2002

- HDTV to take off

Households worldwide:

- On-line   61 million

Percentages below are of on-line households:

- Activated Internet TVs   22%
- Internet through dial-up   78%
- Internet TV viewers with telephone dial-up   60%
- High-speed Internet access   15%
- Internet access through cable modem   12%
- Internet access through digital subscriber lines   6%
- Internet access through wireless connections   3%
- Set-top boxes, information appliances, and Internet game consoles   82 million
- PCs   56 million
- Home network market $4 billion
- Global market for information appliances   56 million
- Subscribers to interactive services   10 million

- Households connecting to the Internet using
  TV-based devices   15 million

Year 2003

Home VOD market providing: $90 billion

- Movies
- TV shows
- Sports
- Other form of VOD entertainment

Year 2004

- Internet appliances shipments to surpass PC shipments
- Digital set-top boxes   252 million
- Revenues from the market $18 billion
- ITV services will generate revenue from:
  - Advertising $11 billion
  - Commerce $7 billion
  - Subscription $2 billion
  - Worldwide digital set-top boxes   252 million

Year 2005

- Internet users   1 billion
- Households to receive satellite TV service   145 million
- Set-top boxes sold worldwide   53 million
- Market for products incorporating $111 billion
  - PC
  - Television
  - Telephone
  - Interactive video transmission function
- DTV homes   222 million
- ITV households   35%
- ITV users   20 million

- Sales of set-top boxes   53 million
- TVs capable of using the Internet   50 million

Development During 2006–2020

Year 2006

- DTV households   20%
- Market of home entertainment combining $109 billion
  - ITV
  - Telephone
  - Computing capability
- Revenues from interactive services $73 billion

Year 2008

- People in developed nations accessing "information superhighway"   80%
- Internet TV viewers with cable TV connection   60%

Year 2009

- Broadband networks (ISDN, ATM, fiber optics) connect the majority of homes and offices
- Electronic banking, including electronics cash, replaces paper

Year 2013

- Market of e-books $66 billion
- Majority of books published on-line

Year 2018

- Market with 50% of all goods sold through information service $208 billion

From research and studies conducted by the following firms:

International Data Corp. (http://www.idc.com)

Inteco Corporation (http://www.inteco.com)

Intel Corporation (http://www.intel.com)

IntelliQuest Research (http://www.intelliquest.com)

Jon Peddie Associates (http://www.jpa.com)

Jupiter Communications (http://www.jup.com)

SRI Consulting (http://www.sriconsulting.com)

Media Metrix (http://www.mediametrix.com)

Strategy Analytics (http://www.strategyanalytics.com)

Find/SVP (http://www.findsvp.com)

Microsoft (dtv) (http://www.microsoft.com/dtv)

WorldGate (http://www.wgate.com)

Motorola (http://www.mot.com)

Consumer Electronics Manufacturers Association (http://www.cemacity
   .org)

Wedbush Morgan Securities (http://www.wedbush.com)

Dataquest (http://www.dataquest.com)

Showtime Networks (http://www.showtimeonline.com)

7

Case Studies

The transition from analog to digital marks a turning point in the history of communications. The most significant new digital communications technologies are the Internet and digital television. The amalgamation of these technologies is leading to the emergence of a fast-growing new industry of ITV. The face of television is changing for millions of viewers. Application and software developers are eager to build on the potential of these new platforms. Programmers and network operators can now plan with confidence for an interactive future.

Approximately 625 million people around the world will have access to on-line services on their TV sets by 2005, including on-line shopping, banking, games, information, and interactive entertainment services, according to research from industry consultants Strategy Analytics. By the end of 2001, the study predicts that 38 million homes worldwide will have access to interactive DTV services, up from 20 million in 2000. Western Europe accounts for 62% of the audience, North America 18%, Asia-Pacific 10%, and Latin America 1%. According to the study, 74% of viewers will use a satellite-based service, 21% cable, and 5% terrestrial.

The most advanced ITV market in the world is the United Kingdom, where 40% of homes will have interactive DTV by the end of 2001. All of the United Kingdom's major digital platforms—satellite, cable, and terrestrial—offer a wide range of interactive services, such as interactive sports coverage, t-commerce, games, e-mail, and walled-garden Internet. Other leading

European markets include Denmark (household penetration by the end of 2001 is estimated to be 25%), Spain (23%) and Sweden (22%).

The case studies discussed in this chapter aim to provide an overview of the state of development of digital broadcasting technologies and ITV services in a few selected regions and countries, such as the United States, Europe, United Kingdom, France, and Australia.

7.1 Case Study: United States

In the United States, satellite and digital terrestrial deployments are advancing as rapidly as technology and public acceptance allows. American trends for DTV are influenced by the fact that 41% of all U.S. TV households now contain a computer. According to the Consumer Electronics Manufacturers Association, about 200,000 DTV products (set-top boxes or integrated receivers) have been sold within the United States since 1998. Of these, 17%, or 34,000 units, are capable of receiving ATSC terrestrial broadcasts, and most of these HDTV products—24,000 units—were sold in 1999. Analysts expect digital terrestrial penetration to reach 50% of all homes by 2006.

7.1.1 Terrestrial DTV

In 1997, the U.S. FCC issued licenses for digital services. The licenses came with implementation targets. The FCC wanted the top four major U.S. TV networks (ABC, CBS, NBC, and Fox) to deliver DTV to the 10 largest U.S. cities by the end of 1998, to all viewers within a few years, and to then withdraw the analog system by 2006. All terrestrial TV stations must be transmitting via digital by 2003. Milestones of DTT implementation in the United States are given in Table 7.1. Major broadcasters have all committed to HDTV, albeit with different digital broadcast formats. The formats being used by various broadcasters are given in Table 7.2.

7.1.2 Cable

Despite its large landmass, cable is very strong in the United States, particularly in urban areas. In the United States, cable is primarily used to deliver TV channels. There are about 68 million households subscribing to cable. There were 5 million digital customers at the end of 1999, with projections for 10 million homes by the end of 2000 and 40 million homes by 2006. In addition, the MMDS wireless cable industry has almost 1 million video customers. Local telephone companies have about 400,000 customers receiving

Table 7.1
Digital Station's Growth in the United States

Date	Number of Digital Stations	Coverage
November 1998	22	33%
May 1999	30	39%
November 1999	41	61%
October 2000	158	—

Table 7.2
Digital Broadcast Formats in the United States

Vertical	Horizontal	Network
720 progressive	1280	ABC, Fox
1080 interlaced	1920	CBS, NBC, PBS, HBO, MSG
1080 interlaced	1280	DirectTV, Primestar, USSB

multichannel video over copper wires using ADSL services. Current rural and metro multichannel subscriber market shares are 58% and 28%, respectively.

The main cable network operators are AT&T, Microsoft, and Time Warner. Over the next 5 years, cable companies plan to invest billions of dollars to convert to high-speed digital networks. At the beginning of 2000, however, Time Warner had only converted 2 million of its 13 million subscribers to digital. Finalizing and implementing U.S. OpenCable hardware and OpenCable Application Platform (OCAP) software standards are a chief cause for the delays in deploying digital cable TV.

7.1.3 Satellite

Satellite services reach 12 million homes, all digital. In late 1999, satellite received a boost when the United States passed legislation giving satellite operators the right to rebroadcast local TV programs. Earlier, consumers had to subscribe to terrestrial or cable to get local programming. In the DBS arena, GM/Hughes' DirecTV has about 7.5 million customers and Echostar's Dish Network has almost 4 million subscribers. These numbers reflect the sub- scriber base before both companies launched interactive OpenTV services. In addition, satellite master antenna television (SMATV) services have 1.4 mil- lion customers. Current rural and metro multichannel subscriber market shares of DBS are 22% and 5%, respectively, and are projected to reach 48% and 11%, respectively, by 2005.

7.1.4   American Viewers

American viewers expect digital TV to provide better quality pictures and sound along with more choices in TV content. The digital TV features most likely to appeal to consumers are better quality sound and pictures (27%), a greater choice of channels (24%), and more choice in types of programming (22%). Indicating general audience preferences, 34% watch news and current affairs programs, 29% sports, 23% comedy, 23% movie channels, and 21% general entertainment channels. Viewers' preferences are based on the 2001 Pace Report, a Gallup survey (conducted October/November 2000) of U.S. adults expressing their views about digital cable TV. The survey was conducted through phone interviews with a national sampling of 1,000 adults.

T-Commerce in the United States

Regarding t-commerce, two-thirds of the U.S. population (based on the Gallup survey, see Section 7.1.4) would consider buying books, CDs, DVDs, vacations, and movie or theater tickets over their computer or televi- sion. They also prefer to shop from home on a TV rather than on a com- puter. Further, 41% have actually purchased goods or services through a TV or computer, with 25% of those using a PC, 10% using a TV, and 6% using both. According to the survey, 27% would prefer to shop at home using their TV rather than their computer. As for types of t-commerce, 75% would be interested in pay-per-view movies and other forms of VOD, 56% in educational programming, 50% in live music events, and 50% in sports events. More women than men said they would prefer to use the TV for home shopping.

7.1.5   Key Players

All of the major companies in broadcasting, telecommunication, cable, and software development have placed stakes in providing digital TV/ITV services.

Tele-Communications

Tele-Communications (TCI), now owned by AT&T, is one of the major U.S. cable companies. It has 14.3 million subscribers and passing 50 million homes (mid-1997 figures). The company is upgrading its analog network to digital using set-top boxes from General Instruments. The upgrading will allow TCI to increase the number of channels it offers from about 40 to 80 where copper wire is installed, and up to 154 where fiber-optic cable is in use. Home shopping and e-mail are planned.

TCI uses the Windows CE operating system in its set-top boxes. The company has also spent $2 billion to buy a stake in News International's *TV Guide* (News International owns BSkyB) to provide the content to TCI's electronic program guides (EPG).

Comcast

Comcast was the fourth largest cable company in the United States in 1999. Comcast has a 57% stake in QVC, the large home shopping network. Microsoft has a $1 billion (11.5%) stake in Comcast.

AT&T

AT&T is now the United State's largest cable company with 16 million customers. After the TCI merger in May 1999, AT&T formed an alliance with Microsoft. This allowed AT&T to take over cable operator MediaOne for $57 billion. MediaOne owns part of the United Kingdom's One2One mobile telephone company and 29.9% of Telewest Communications, the United Kingdom's No. 2 cable company. AT&T already has a 21% stake in Telewest Communications through its TCI acquisition. As a result of an alliance with Microsoft, AT&T provides Microsoft's Windows CE operating system in its set-top decoders.

Microsoft

Microsoft has taken a $5 billion stake in AT&T as part of the company's goal of reaching the TV masses and being the gatekeeper of all electronic mediums. Microsoft also owns WebTV. WebTV boxes cost $300, with a monthly subscription of $19.95. WebTV also uses the *TV Guide*. In the United Kingdom, Microsoft has teamed up with British Telecom (BT) and the BBC for a trial of WebTV. It also has a $500 million stake in NTL, the United Kingdom's major cable operator, and a stake in Telewest Communications.

Sun

Sun Microsystems provides a number of software programs based on Sun's Java, including Java API and ITV receivers (Section 3.6). Sony, with its Perios operating system, has teamed up with Sun Microsystems and its Java software.

AOL

AOL is the world's biggest ISP with 29 million subscribers in 15 countries. This includes both AOL and Compuserve users. In 1997, it bought NetChannel. In January 2000, AOL and Satellite Company DirecTV announced a WebTV offering using set-top boxes from Philips Electronics and Hughes Network Systems. Later versions of the set-top boxes could be using technology from TiVo. The aim was to get a big slice of the 30 million U.S. households predicted to be using ITV by 2004.

On Jan. 10, 2000, AOL and Time Warner announced a $350 billion merger, with AOL paying $184 billion for a 55% stake in the new company. The old one-way media of print and TV gave way to the new interactive Internet media. With this merger, AOL has access to Time Warner's U.S. cable network with 12.6 million customers, as well as marketing. The combined group sees three revenue streams (with current revenue share in parentheses): subscriptions to the Internet and cable (40%); advertising and e-commerce (20%); and content (40%). In Europe, *AOL Europe* is a joint venture with the German company Bertelsmann, and in South America, *AOL Latin America* is a joint venture with Cisneros. Just a few days after the AOL-Time Warner merger, Warner Music made a bid for EMI to create the world's biggest music business valued at £12 billion.

WINfirst

WINfirst is a fiber-optic network company competing with U.S. cable TV operators to build high-speed digital systems. In February 2001, Canal+ Technologies, a unit of Vivendi Universal, announced a deal with WINfirst to deliver ITV services to homes. WINfirst will offer Canal+ technology, interactive digital TV services, and original content to more than 3.7 million cable-connected households. In addition, WINfirst will also deploy Canal+'s MediaHighway software and MediaGuard access systems (Section 7.2) in multiple key markets across the United States.

More than 20 digital operators and broadcasters have deployed Canal+'s MediaHighway and MediaGuard products in France, the United

Kingdom, Asia, and the Middle East. Its software runs on more than 9 million digital set-top boxes worldwide.

News Corp.

Murdoch's News Corp. has been making numerous investments into set-top box software companies, such as OpenTV and ACTV, and also in TV channels. News Corp. already has a minority stake in TheStreet.com, an on-line financial newspaper, as well as other health care companies.

Disney

Disney owns 10 TV stations, the ABC network, 42 radio stations, and 9 international Disney TV channels. It also owns Internet portals, a library of films, TV programs, and has interests in 9 U.S. cable companies.

DirectTV

Satellite broadcaster DirectTV is owned by Huges Electronics. It has acquired USSB and Primestar, increasing its reach to about 9 million subscribers. It purchased Vignette's StoryServer 4 software for its Web site so that customers can personalize their TV program information. DirectTV is providing an in-flight TV service to JetBlue, America's latest budget airline. Passengers can pay to access 24 live TV channels.

EchoStar

EchoStar is a satellite broadcaster with 300 channels of digital video and CD-quality audio delivered to 4 million customers. One of its channels is MediaX, which has an Internet site as well as other features, such as rich interactive multimedia, on-line purchasing, and live events. The e-commerce site offers 265,000 entertainment titles, including audio CDs, videos, and DVDs. These features are delivered to both PCs and TVs. The site is popular with young, tech-savvy audiences.

iSKY

iSKY is a new start-up that plans to launch two-way Ka-band satellite services by 2002. Starting with broadband Internet before moving into ITV, iSKY plans to deploy three satellites covering North, Central, and South America (Section 6.4.4).

USA/Lycos Interactive Networks

USA/Lycos Interactive Networks was created with the merger of USA Networks with Lycos in 1999. The company owns the Home Shopping

Network (HSN) and has controlling rights in Ticketmaster. The company aims to produce synergy between the Internet and broadcast TV.

Companies in the Business of Set-Top Boxes

Time-Warner is working on a set-top box that will provide both VOD and Internet access. In May 1999, Sun Microsystems formed an alliance with AOL to provide software for set-top boxes. Power-TV and @Home (a cable Internet provider with 3 million subscribers and now part of the $7.5 billion merger between Excite and @Home) are also working to design set-top boxes. Liberty Digital has made many investments in set-top box software companies and TV channels.

    Note: Appendix 4.B lists other enhanced and ITV developers, producers, and consultants.

7.2   ITV in Europe

The United Kingdom and Western Europe currently lead the world in ITV subscribers. Interactive TV by satellite, cable, and terrestrial broadcasting is booming in the United Kingdom and Western Europe. All of the digital TV networks launching in Europe are interactive. OpenTV's set-top boxes are installed in 13.9 million homes. An analog closure date of 2010 has been proposed for Sweden, the Netherlands, Spain, and Germany. The heavily cabled countries, such as the Netherlands, Sweden, and Germany, will be able to switch off analog, since only a tiny proportion of viewers are actually dependent on terrestrial service. The setting of the analog end date by these countries will influence customer behavior.

    European countries have adopted the digital video broadcasting (DVB) standard. DTV was first introduced in Europe via satellite. Evidence suggests that no European country can support more than one satellite platform. France has been the notable exception (Table 7.3). Satellite operators are looking to high-speed Internet access for future revenue models. Public service and free-to-air broadcasters are the driving forces behind DTT's expansion in Europe. The DTV services available in Europe through various delivery systems are provided in Table 7.4. In this table, a country is included if either digital satellite services are broadcast from that country or there is an agreement to broadcast over that country from another country. For example, in Luxembourg and Belgium, there are no digital broadcasts but satellite services can be received from neighboring countries.

Table 7.3

Satellite Services in Europe

Country	Operator	Satellite	Remarks
United Kingdom	BSkyB	Astra[1]	Launched in late 1998
France	Canal/TPS	Astra/Eutelsat[2]	Over one million subscribers
Spain	Canal Satellite Digital/ Via Digital	Astra/Hispasat	Hispasat is the remaining national satellite project
Germany	DF-1	Astra	—
Italy	—	Eutelsat	Satellite penetration slow. Large number of free-to-air terrestrial channels.

1. Orbital slot - 28.2°E
2. Orbital slot - 29°E

A concern for regulators is allowing the DTV market to grow without abusing various provisions by competitors. Europe has a *Treaty of Rome* to allow healthy development of digital TV in the region. Some areas of concern and how they might be addressed are depicted in Table 7.5. A major area of concern is the use of conditional access (CA) by various providers. In Europe, a number of CA systems are in use (Figure 7.1). Article 86 of the Treaty of Rome provides for two types of CA systems, simulcrypt and multicrypt.

ITV has already penetrated in Europe, especially in the United Kingdom and France. Canal+ leads the world with more than 4 million ITV subscribers among its 13 million digital satellite and cable customers across Western Europe. If U.K. numbers are added to this, there are approximately 6 million actual ITV users in Europe. Over the next five years, ITV penetration in Europe is expected to grow from between 15% and 20% today to 60% and, eventually, 90%.

The API, the software operating system for the set-top box, controls the types of interactivity. A number of different systems are available in Europe. Details on these are offered in Table 7.6.

We shall examine the developments of DTV and ITV in the United Kingdom and France now.

Table 7.4
DTV Services in Europe

Country	Satellite	Cable	Terrestrial
Austria	X	X	—
Belgium	*	X	—
Denmark	X	X	—
Finland	X	—	—
France	X	X	—
Germany	X	X	—
Greece	—	—	—
Ireland	X	—	—
Italy	X	—	X
Luxembourg	*	X	—
The Netherlands	X	—	—
Portugal	X	—	—
Spain	X	X	—
Sweden	X	X	X
United Kingdom	X	—	X

* Satellite services received from neighboring countries. (*From:* IDATE.)

7.2.1 United Kingdom

More than 3 million households in the United Kingdom have access to digital TV, and this is expected to rise quickly. As per projections, there will be more than 8 million digital subscribers by 2003 and 18.5 million by 2008. Broadcasters hope that the growth in digital will mean that they can switch off the analog TV signal by 2010. The £22 (≅ 33 billion auction of bandwidth to facilitate G3 telecommunications promises that the analog frequencies released by the switch to digital should fetch even more. The money obtained will encourage the government to find a solution to the problem of legacy analog sets.

Despite DTV starting rather late in the United Kingdom compared with some other European countries, rapid developments during the latter

Table 7.5
Area of Concern and Treatment by Treaty of Rome

Area of Concern	Article No. of Treaty of Rome	Objective of the Article	Remarks
Alliances	85	Prevents agreements between enterprises that have the object or effect of distorting, preventing or restricting competition within the EU	The Commission has blocked four separate attempts to create alliances through which to launch a digital television platform
Conditional Access	86	Prevents companies from abusing their dominant position in any market-obligation to supply conditional access on fair, reasonable and nondiscriminatory terms	Principle incorporated in the Television Standard Directive: simulcrypt and mulicrypt approaches adopted
Electronic Program Guides	86	No specific regulation of EPGs has been enacted on a European level	—
Application Program Interface (API)	86	—	—

part of 1998 and 1999 have made it the most promising country in Europe for interactive digital services. A range of competitive technologies has been tested for the provision of ITV services in the United Kingdom. The U.S.-based WebTV Networks, Inc. has been working with BT since March 1998 to deliver interactive services through Pace Micro Technology's set-top boxes, over copper wire, to standard analog TVs. Internet and multimedia e-mail access is part of this package. The use of VideoFlash™ technology, allowing the downloading of full-screen, full-motion, high-quality video, graphics, voice, and music to standard TVs is also included in this package. In 1998, Kingston launched the country's first commercial ADSL service. BT also ran an ADSL trial to deliver interactive services to consumers over copper wire in London during the latter part of 1998 and the first quarter of

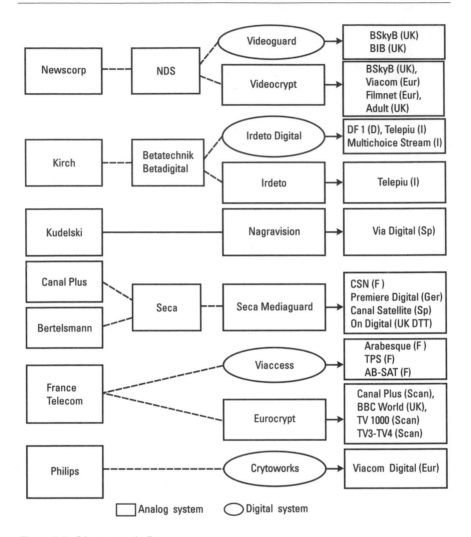

Figure 7.1 CA systems in Europe.

1999. BT is in the process of deploying high-speed broadband networks to compete with the cable companies. Its BeTaNet is an advanced Internet/multimedia system based on terrestrial fiber networks. BT is collaborating with other European telecom companies to create the largest pan-European high-speed network. Both BeTaNet and the European network are based on synchronous digital hierarchy (SDH) and dense wave division multiplexing (DWDM). When integrated with ASDL, BeTaNet will

Table 7.6
API Systems Used in Europe

Systems	Designers
OpenTV	Thomson Multimédia
MediaHighway	Seca
Sun Chorus	Sun Microsystems
Power TV	Scientific Atlanta
DTV Navigator	Network Computer, Inc.

From: IDATE.

bring interactive digital broadband communications within reach of most residential telephone customers in the United Kingdom.

March 2000 saw the launch of the first nationwide ITV advertisement in the United Kingdom on the Open platform backed by BSkyB, the British satellite broadcaster. Cable TV provider Telewest has started offering a full range of ITV services. The U.K. ITV market is now one of the most advanced markets for ITV in the world.

However, the United Kingdom also has no standard platform for delivering ITV. There are four main service providers: ONdigital, which uses a terrestrial broadcast network (ONdigital has decided to change its name to ITV Digital in order to gain a higher consumer profile using the brand of the United Kingdom's largest commercial broadcaster); Open, which broadcasts via the BSkyB satellite; and cable providers NTL and Telewest. ONdigital has an MHEG-based system, the two cable providers both base their services on HTML and JavaScript, and Open has its own proprietary technology.

7.2.1.1 Digital Satellite

The United Kingdom's first commercial DTV service started in October 1998 with the launch of SkyDigital. The DBS service from BSkyB is provided to U.K. homes from the Astra satellite. Approximately 551,000 SkyDigital customers get cheap on-line access to the Internet, as well as discounted telephone services and 140 digital channels. The company has announced plans to switch off its analog service by December 31, 2002.

7.2.1.2 Digital Terrestrial

BSkyB's main rival, OnDigital, broadcasts 30 digital channels through digital terrestrial to rooftop aerials. The company expects to reach 170,000 subscribers by 2003–2004. ONdigital had been seen as an important driver for the rollout of DTV in the United Kingdom. Apart from ONdigital, the BBC and other free-to-air analog terrestrial broadcasters have all planned or launched digital services, having been allocated the three digital multiplexes with the highest national coverage (90%) by regulators. A further three multiplexes, with lower national coverage of around 80%, have been allocated to new pay-TV services. The BBC is carrying BBC1, BBC2, BBC News24, and BBC Choice on its multiplex. The ITV group and Channel 4 group have also announced digital offerings from their shared multiplex. Other regional free-to-air broadcasters, including the Welsh channel S4C and some Gaelic services, have taken up their entitlement on the third-largest coverage terrestrial multiplex.

7.2.1.3 Digital Cable

The United Kingdom was a late starter in the cable industry, compared with most of the neighboring European countries. One advantage of this late start is that the cable networks are capable of supporting digital services. From 1996–1998, as a result of consolidation, the number of significant providers fell from 24 to just 5. Industry analysts suggest that the U.K. cable industry will finally consolidate with only one provider, although industry regulator and consumer protection groups may prohibit this. Current leaders in the U.K. cable industry are Telewest Communications, Cable & Wireless Communications, and NTL. The main strategy pursued by these companies has been to provide Internet capability with their set-top boxes. There is no agreement on a unified platform.

ONdigital and Open will benefit from the widespread introduction of ADSL lines by U.K. telecommunications providers. These will reach 80% of the U.K. population within three years. They offer broadcasters a broadband line to homes independent of the existing cable network, giving them the potential to provide significantly improved levels of interactivity.

7.2.1.4 U.K. Consumers

The United Kingdom is being driven by high levels of competition for ITV services over cable, satellite, digital terrestrial, DSL, and even wireless. Because of aggressive marketing and improved interactive content, more people who do not have digital TV are now thinking seriously about getting it. The 2001 Pace survey provides interesting input about consumers' preferences in the United Kingdom. It shows that U.K. consumers want better quality sound and pictures and they value greater choice in channels and programming, even if they do not always take advantage of all their choices (i.e., mostly watching just 10 channels out of the hundreds available). About 28% of consumers already receive digital TV services. This figure is more than double the 13% statistic from the 2000 survey. One percent of U.K. households receive no TV at all. Reasons cited for adopting digital TV, apart from the Crown switching off analog TV, are very interesting. For example, 51% cited better quality sound and pictures, 43% cited more movie channels, and 40% cited a greater choice in types of programming.

About a quarter of the respondents (23%) reported having actually used either a computer or DTV to buy goods or services, with half of them buying vacations or holidays ahead of any merchandise. Among this group, 77% have done transactions by computer and 23% by TV, which includes teletext. According to the survey, 37% prefer using a computer for home shopping, 32% prefer a DTV, and 22% prefer neither for their shopping.

A year ago, in 2000, 23% reported accessing the Internet at home. A year later, the number jumped to 42%. Of these, only 6% used a DTV for Internet access. (Internet access on a DTV has been available in the United Kingdom for less than a year.)

As for types of t-commerce, 71% like pay-per-view movies, while 57% are willing to pay for news and current affairs programming, which is more than those willing to pay for music or sports. Suggesting cultural preferences, 43% watch news and current affairs, 42% movies, 36% sports, 36% general entertainment channels, and 33% watch music and arts.

7.2.1.5 Key Players

NTL

NTL, a U.S.-backed company, offers Internet technologies, broadband networking, and digital video delivery. NTL's fiber-optic broadband network passes 5.6 million homes in the United Kingdom. NTL's current customer base, combined with the recent acquisitions of Cable & Wireless' cable assets and Switzerland's Cablecom, reaches 4.2 million subscribers. NTL has businesses in France, Australia, and Ireland. NTL provides movies, TV programs, home shopping, music, news, education, and travel services.

Cable & Wireless

Cable & Wireless is a U.K.-based company with an international asset base, of which Hong Kong Telecom is the most significant component. Cable &

Wireless Communications services passed 4.286 million U.K. homes and had 660,000 dual telephone and TV subscribers. In 1998, Cable & Wireless Communications bought a controlling share in Two Way TV, a media company that will provide interactive game shows as part of its "TV Mall" offering.

Open

Open broadcasts via the BSkyB satellite and local telephone lines, which are comparatively old and slow. Open was the first broadcaster to offer a national interactive service. It quickly demonstrated how great the potential for ITV will be. Though the service was only available through untested technology in a limited number of homes, purchases through Open quickly hit £1 (≈$1.5) million per week. Open now provides shopping, banking, and e-mail services to 2.8 million subscribers.

Telewest

Telewest is a U.S.-backed company. In May 1999, Telewest services passed 4.2 million U.K. homes and had 800,000 dual telephone and TV subscribers. Through its proposed service Active Digital, Telewest plans to offer e-mail, home banking, shopping, and interactive entertainment via the TV. These services will be available to all Telewest's 110,000 digital subscribers, helping it increase its subscriber base to a predicted 500,000 by the end of 2001.

Microsoft

Digital services via cable networks got a boost in May 1999 when Microsoft took significant stakes in both NTL and Telewest. As part of its alliance with AT&T, Microsoft announced that it was committed to buying the 29.9% stake in Telewest, then held by U.S. cable operator MediaOne. Microsoft already holds a 5% stake in NTL.

Pace

Pace is the leading manufacturer of set-top boxes. ONdigital interactive terrestrial service in the United Kingdom has 552,000 Pace set-top boxes in homes, using the Canal+ MediaHighway and MediaGuard CA systems for free and pay ITV services. News Corp.'s BskyB has deployed 2.6 million SkyDigital set-top boxes from Pace in the United Kingdom. Cable & Wireless has deployed more than 60,000 digital Pace boxes enabled for interactive cable services using the Liberate ITV system based on Internet standards.

7.2.1.6   ITV Content

SkyDigital has collaborated with content providers such as British Interactive Broadcasting (BIB). BIB was formed in May 1997 to deliver digital ITV services and is owned jointly by BSkyB (32.5%), BT (32.5%), Midland Bank (20%), and Matsushita (15%). BIB's service partners include retail companies such as Great Universal Stores, Iceland, Kingfisher, and Woolworth, as well as companies such as Midland Bank, Ford, Unilever, and Coca-Cola.

Sky Online, an independent entity of SkyDigital, will develop a range of Internet-based digital interactive services linked to sports, documentary, quiz, and other entertainment programs. It also plans to offer learning services.

Interactive education will have a significant role in the development of ITV. BBC launched its BBC Knowledge channel on June 1, 1999. BBC Knowledge encompasses a range of programs for children, parents, teachers, and other learners at home. User interaction with BBC Knowledge is initially through the Web but ultimately via the TV screen itself. Wales Digital College is also due to start a number of innovative ITV trials leading to full service in Wales in conjunction with the Welsh Medium Channel S4C. The Open University also has many programs on interactive learning.

7.2.1.7   Standards Dilemma

The standardization issue is yet to be addressed (Section 3.7). As a result, any advertisement designed to run on all networks in the United Kingdom needs to be remade several times in order to work on all of the different systems. Subscribers are also confused by the different offerings.

7.2.1.8   Commercials

Advertising is crucial to ITV's future. Forrester Research predicts that the total U.K. ITV market would be worth U.S.$20 billion by 2004, of which $2 billion would be subscriptions, $7 billion commerce, and more than half—$11 billion—would be advertising revenue.

Unilever's Commercial: "Chicken Tonight Sauces"

Unilever's first ITV advertisement in the United Kingdom aired in March 2000 on BSkyB's Open service, representing an important milestone. The ad, for Chicken Tonight sauces, offered consumers the opportunity to receive a recipe book and a money-off voucher with the option to see recipes on Open. Consumers used the remote control to visit the company's

Creative Kitchen site, apply for cash-back vouchers and, in some areas, order on-line by accessing the TV sites of nearby grocery stores. Feedback has been positive, and similar projects are in the works with other advertisers. This shows the potential of one of the key strengths of ITV technology for advertisers—the fact that all the data on viewers is already held in the system. The viewer simply clicks on an icon to receive the on-screen offer.

7.2.1.9 T-Commerce

ITV is expected to surpass the Internet as Europe's primary e-commerce platform by 2005, perhaps because security fears about Web shopping could abate in the safer, more familiar TV environment. Tropical Places, the first company to sell holidays on British ITV, announced in January 2001 that it had taken more bookings through t-commerce than through its Web site.

T-banking will enable consumers to check their accounts and pay bills. It will also arrange mortgages, life insurance, and personal loans. Datamonitor's survey of 35 leading financial service institutions in the United Kingdom suggests that almost 75% of retail banks either offer t-banking already or are negotiating with broadcasters to develop such a service.

7.2.2 France

France has been an early adopter of DTV services, mainly via satellite. The country also has the highest number of households subscribing to such services, compared with other countries within Europe. DTV currently penetrates 12% of French households. In March 2000, out of 22.58 million households, there were 2.68 million digital households. This represents more than 60% of the 4.265 million multichannel (satellite or cable) households. There are 417,000 digital cable homes (out of 2 million cable homes). Canal Satellite broadcasts to 1.415 million homes and Télévision Par Satellite (TPS) broadcasts to 850,000 homes.

7.2.2.1 Satellite

Canal Satellite, a subsidiary of Canal+, launched the first digital TV service in 1996. The development of digital TV in France was essentially due to satellite TV. Three were launched in 1996: Canal Satellite (April), TPS (December), and AB Satellite (December). AB Satellite disappeared as an operator and those channels are now broadcast via Canal Satellite. With close to a million subscribers, Canal Satellite is twice as big as its national rival, TPS. Canal Satellite transmits from the SES/Astra satellite and can be received using a digital set-top box with MediaGuard conditional access and MediaHighway API technology developed within the company. TPS transmits from Eutelsat's Hotbird satellite and uses the conditional access Viaccess, developed by France Télécom, and API technology from OpenTV. Two different versions of AB Satellite are broadcast, both on Eutelsat and Astra using Viaccess. In addition, simulcrypt agreements have been negotiated with its two competitors.

7.2.2.2 Cable

Cable operators are all currently converting their analog households into digital households. Among those already providing digital TV access are three major operators: Lyonnaise Câble, which owns 57.7% of digital cable households; NC Numericable (a Canal+ affiliate), which owns 24% of digital cable households; and France Telecom Cable, which owns 17.5% of digital cable households.

7.2.2.3 Key Players

TPS

TPS is a joint venture between free-to-air terrestrial broadcasters France Télécom (the dominant telecommunications operator and one of the leading cable operators) and Lyonnaise Câble, which is also a major cable operator. TPS is owned by TF1 Group (25%), M6 Group (25%), Sues Lyonnaise des Eaux Group (apart from M6) (25%), France Telecom Group (16.5%), and France Television Group (8.5%).

Canal Satellite

Canal Satellite is owned by Canal+ (66%) and Lagardere (34%).

7.2.2.4 ITV

In France, ITV started in 1997. In recent years, the technology is evolving well and the service applications have been well accepted by consumers, advertisers, and broadcasters. The key players are addressing four media marketing objectives: increased traffic (advertising impressions); increased brand awareness; database development; and transactions and sales. To achieve these objectives, the players are providing personalized services and relation-based advertising that offers a new type of relationship between the brand and the consumer. A number of innovative services are being provided to customers.

Canal Satellite/TPS Services

Canal Satellite and TPS provide the following ITV services to their customers:

Program Guides Canal Satellite uses the *Mosaique* EPG, which offers information and selection services about programs proposed on different channels.

TPS uses its *Guide des programmes*, which has information, navigation, and selection services for programs broadcast on TPS. The guide has links with an information page associated with each channel and with other services.

Pay TV Canal Satellite has its *Kiosque* pay-TV services for movies, sports, and special events, including a "multicamera" capability for zapping. TPS has its *Multivision* pay-TV services for cinema, sports, special events, and theatre, including a multicamera capability for zapping. The business model involves a monthly subscription.

On-line Shopping Canal Satellite has a *Forum des boutiques*, a transaction gateway leading to 16 services. There are stores and thematic channels based on merchandising in addition to a personalized marketing capability. TPS offers *Galerie TPS boutique*, which has retail capabilities and an information area, including temporary or permanent advertising material.

Interactive Content Both services have information connected to linear channels. Interactive content offered by Canal Satellite includes the following:

- General information relating to weather, job offers and employment, finances, and topics of interest to people;
- Local information;
- Games: *Canalsat Jeux*—between 4 to 6 different games, which are often changed; *Piktorézo*—a network game;
- Background sports information and statistics;
- T-business services, home banking;
- Betting services;
- Travel and tourism in association with *Liberty* channel;
- E-mail service.

TPS provides the following interactive content through its portals:

- *Info Express*: information portal about news, culture, and sports;

- *Meteo Express:* weather service on demand;
- *Infoscore:* information about sports, events, and results;
- *Bandiagara:* a game portal that includes nine memory games, which are often changed;
- *Espace annonces:* small job, car, and property ads;
- *Fi* financial services—information and transactions in association with banks;
- *PMU:* bet-at-home service;
- *TPS le mail:* e-mail service.

7.2.2.5 Commercials

A number of interactive advertising campaigns have been launched. TPS and Canal Satellite have developed their offers through different technology (MediaHighway for Canal Satellite and OpenTV for TPS).

Both Canal Satellite and TPS are pursuing the same business model of using databases and infomercials to create increased traffic. They are also trying to personalize the different services.

Adecco: Direct Marketing to Job Seekers (TPS)

This campaign meant for job seekers ran March 8–21, 1999, on the TF1 and M6 channels, plus Galerie TPS boutique. Job seekers could browse job opportunities, get information, and arrange an interview based on time slots available. In the end, 4,019 applicants called and nearly 300 received job offers.

7.3 Australia

In Australia, the pace of development of digital TV, as well as ITV, is slow, albeit the government has fixed 2010 as the target for analog switch off. The country's free-to-air TV market may soon become interactive.

7.3.1 ITV in Australia

There are more than 11 million TV sets in Australia, with multiple users in each house. Australia introduced digital TV on January 1, 2000. A late starter, but the projections are that within two years, half the Australian population will be using digital TV. Australia's first ITV was launched in

November 2000. It is expected that 50% of Australians will be using interactive services regularly after the full deployment of interactive solutions by free-to-air channels. ITV will become widely available in metropolitan areas in the first half of 2001. The set-top boxes are expected to cost around $300, but it is possible that pay-TV networks may incorporate this cost into their subscription, as BSkyB has done in the United Kingdom.

7.3.2   Key Players

7.3.2.1   Ice Interactive

U.S. computer giant Oracle Corporation established Ice in February 2000. Oracle owns 15% of Ice; Liberate Technologies in the United States owns 10%. The Australian-based technology investor Burdekin Pacific provided local venture capital funding, and it is the majority shareholder with 40%. Ice is planning to launch ITV trials in New Zealand in 2001.

7.3.2.2   Austar United Communications

The Australian regional pay-TV operator Austar United Communications is already offering an ITV service to subscribers. Its weather channel shows interactive ads for Toyota, Vodafone, and Wesfarmers Dalgety. Austar is replacing subscribers' set-top boxes with new boxes that have a computer memory and allow a telephone connection.

7.3.2.3   Foxtel and Optus

Foxtel and Cable & Wireless' Optus are still planning their move into ITV services. To offer T-commerce, they need to replace their existing set-top boxes, and Optus will need to upgrade its cable network from analog to digital. Optus has announced that it will spend $1 billion in preparation for ITV.

7.3.3   Pizza Hut T-Commerce

Pizza Hut paid $50,000 to join Australia's first t-commerce trial. TV software company Ice Interactive conducted the trial on regional television station WIN in November 2000. The trial was conducted on 150 houses. By aiming a specially designed remote control at the Pizza Hut icon on the screen during a Pizza Hut ad, viewers can select from 16 types of pizza, plus drinks and garlic bread. A set-top transmitter sends the order through the telephone line. The order is delivered as an e-mail at the local Pizza Hut restaurant. Payment is taken manually when the order is delivered. Using the keyboard, people can also send and receive e-mails through their TV sets.

7.4 Lessons Learned from the Case Studies

7.4.1 The World Is Going Digital

Digital provides more channels, high-quality pictures, endless reproductions with no degradation, and interactivity. The shear force of technology will revolutionize the world. All countries are following the digital path.

7.4.2 Business Is Changing

Technology is transforming the world of business, and it is increasingly clear that no enterprise, new or long established, large or small, will be immune from the impact. Digital changes all the rules, internally and externally, be it network operations, content creation, service distribution, or marketing and customer relationship management. From basic human resources through to the final cost structure of the service, the business systems profile is becoming dramatically different.

7.4.3 Far-Reaching Impact of T-Commerce

Transactions are at the core of any business operation, forming one of the most significant components of business systems and overall costs. All transactions are information-based, relying on exchanging, obtaining, using, and tracking information throughout their life cycle. Hence, t-commerce shall have far-reaching implications on all businesses. T-commerce is ideal for selling impulse items and low-priced products. As with e-commerce, consumers are often wary of buying expensive items through the t-commerce medium. And t-commerce gives television networks the opportunity to earn more from advertising because higher rates can be charged for "t-enabled" ads.

7.4.4 Radical Structural Changes Are Required to Face the Change

Digitalization will ultimately affect (and in some aspects is already affecting) all businesses of broadcasting and all facets of those businesses. Established companies are facing and undergoing radical structural and system changes to maintain their competitiveness and take advantage of the new opportunities that t-commerce brings.

New companies are emerging in a very different environment from their predecessors. A new business can enter a broader range of markets faster and more effectively than ever before. Information processing and networked operations lower the traditional barriers to market entry and facilitate global access. However, enhanced opportunities also mean an emerging business has to be prepared to meet the demands of those opportunities immediately and effectively. Traditional patterns of growth and development no longer apply in many cases.

7.4.5   Collaboration and Mergers Key to Survival

Collaboration, mergers, and partnerships between various players will help bring together the expertise of all the operators. Conglomerates providing TV broadcasting, film, and TV production, consumer publications, and other entertainment services will survive. As the digital infrastructure becomes established, content will become the battleground to gain and retain customers. The customer is and will remain the king.

Selected Bibliography

http://australianit.news.com.au/common/storyPage/0,3811,1869277%5E442,00.html, April 2001.

http://www.brw.com.au/, December 2001.

"Hughes Network Systems," http://www.hns.com/, December 2001.

Merle, I., Carat France, imerle@carat.

Freed, K., *Multichannel News International*, February 2001.

Priestley, M., Carat, London, http://www.itvnews.com/intnews/index.htm, April 2001.

Hannen, M., "T-commerce: Ads with the Lot," *BRW*, Vol. 22 No. 45.

Glossary

AA   average audience (a measure of TV viewership)

AAES   American Association of Engineering Societies

ABES   Association for Broadcast Engineering Standards

ABR   (available bit rate) ATM service that adjusts the rate of data transmission to try and accommodate the transmitting device

AC-3   a standard encoding method, Dolby AC3, using an algorithm to compress TV audio signals

Access control   a collection of procedures put in place by service providers to regulate access to their content. For example, a service provider might let only authorized users watch pay-TV programs.

Access provider   see Internet service provider

Active server pages   (ASP) ASP script extensions contain either Visual Basic or JScript code. When a browser requests an ASP page, the Web server generates an HTML page according to the ASP code and sends it back to the browser. ASP pages are similar to CGI scripts, but they enable Visual Basic programmers to work with familiar tools. This is a page that performs customized applications services.

ACTS (Advanced Communications Technologies and Services—European Commission) program for R&D supporting many technology projects, including ITV-related technologies

ADC term used for analog-to-digital converter, which is used to convert analog signal to digital signal

Ad-click rate sometimes referred to as "click-through," this is the percentage of ad views that resulted in an ad click

Ad clicks number of times users click on a Web ad banner

Ad views sometimes called "impressions." Number of times a Web ad banner is downloaded and presumably seen by visitors. If the same ad appears on multiple pages simultaneously, this statistic may understate the number of ad impressions, due to browser caching. Corresponds to net impressions in traditional media. There is currently no way of knowing if an ad was actually loaded. Most servers record an ad as served even if it was not.

ADN (advanced digital network) usually a 56-Kbps leased line

ADPCM (Adaptive Differential Pulse Code Modulation) similar to PCM encoding except the encoding is the difference between the sample amplitude and the previous sample

ADSI (analog display services interface) Bellcore standard for data and voice over conventional phone line. Used for advanced telephone services. Could be used for shopping and banking to home terminals.

ADSL (asymmetric digital subscriber line) see DSL

ADTT (advanced digital television technologies) a pan-European research program partly replacing Eureka 95 (HD-MAC project). The ADTT covers digital TV production, transmission reception, and display equipment.

AEA American Electronics Association

Aesthetic   refers to factors in content, packaging, or presentation that are not economic or technical

Algorithm   specification of the steps that perform a process, such as drawing a circle or compressing an image

ANSI   (American National Standards Institute) a standards organization in the United States

Analog   transmission (of audio-video) by a continually variable waveform signal

Analog set-top box   a device for linking standard TV sets to the interactive TV network. Analog set-top boxes receive and unscramble analog transmissions for TVs. New generations of advanced analog set-top boxes also allow for some data services, such as EPGs, to be encoded in the same radio spectrum as the video signal.

API   (application programming interface) a series of functions that programs can use to make the operating system do their work. APIs allow hardware manufacturers and independent third parties to develop applications that can interface with the operating system. API is a built-in function of the set-top box that forms the link between its internal processor and the outer layer of the user interface, such as the EPG.

Applet   a small program embedded in an HTML page and run by the user's browser. Most commonly a Java program. Applets differ from full Java applications, as they cannot access resources of the computer such as files and network devices, except to the computer they were downloaded from. This is one way to deliver local applications or service-based functionality to an ITV terminal.

ASP   (application service provider) Internet providers that allow a company to lease software on their servers and run the software without having to own them. See also active server pages.

Aspect ratio   the ratio of TV picture width to height. In PAL and NTSC video, the present standard is 4 by 3 (or 4:3). For HDTV, the standard is 16 by 9 (or 16:9).

Assets   actual audio, video, and image data that are referenced by a project. The term asset is synonymous with resource when used at runtime.

Asynchronous   a method of communications where data packets are sent when the channel is available. This causes variable transmission delays, which are not very suitable for real-time video or audio.

Asynchronous transfer mode   (ATM) emerging standard for very high-speed switching and routing of integrated data, voice, and video over broadband networks. ATM is a high-bandwidth, low-delay, packet switching and multiplexing technique based on a fixed-length, 53-byte cell. All broadband transmissions (video, audio, text, stills) are divided into a series of cells and routed across an ATM network consisting of links connected to ATM switches. ATM uses a system of creating and maintaining virtual point-to-point connections. Its speed and scalability make it the most attractive technology for broadband communications. The most significant feature of ATM is its uniform handling of services, allowing one network to meet the needs of many broadband services. ATM is projected as the most suitable technology for advanced ITV systems. It is finding commercial application in backbone data networks.

ATSC   (Advanced Television Systems Committee) a committee formed in the United States to establish technical standards for advanced TV systems, including digital HDTV

ATVEF   (Advanced Television Enhancement Forum) a cross-industry alliance of companies representing the broadcast and cable networks, TV transports, consumer electronics, and PC industries. This alliance has defined protocols for HTML-based enhanced TV, which allows content creators to deliver enhanced programming over all forms of transport (analog, digital, cable, and satellite) to any intelligent receiver.

Audience flow   shift of audience away from one broadcast station to another with the change of program, as compared with those that remain with the original station

Audio monitor   speakers for listening to audio

Authentication   the process that checks that an application is authentic and, in doing so, sets the ID of the application. The ID is later used to determine the application's permissions regarding privileged functions it may access and permissions in the file system.

Authoring   developing (writing of text, recording of audio, importing of video, inserting graphics) hypertext and hypermedia learning, entertainment, and reference materials. The term "authoring software" is often confused with "presentation software" designed for development of graphics and, possibly, some multimedia components. The latter is similar but with less capability.

AVI   (audio-video interleave) Microsoft's format for multimedia files and video capturing

B-frame   (bidirectional predicted frame) part of the MPEG standards, B-frames track changes from the preceding and following reference frames in a group of pictures

Banner ad   an advertisement box on a Web page that is "hot-linked" to the advertiser's site. Banner ads carried on Web sites offer viewers the chance to "click through" to see information the advertiser wishes to make known to the viewer. Generally, this links to the advertiser's main Web site. Unlike TV or print ads, banner ads offer an extra dimension to advertising—the advertiser's site can become a direct marketing channel. It also gives vendors of advertising space another way to calculate the value of their site. The concept is being extended to ITV, where conventional ads can be selected through "click through," or banners can be displayed continuously through the program.

Backbone   a high-speed line or series of connections that forms a large pathway within a network. The term is relative to the size of network it is serving. A backbone in a small network is much smaller than many nonbackbone lines in a large network.

Bandwidth   range of frequency spectrum available or needed to transmit data (pictures, sound, digital packets) over a medium, such as a cable or air, or through an electronic device. The higher the bandwidth available, the greater the amount of data that can be transferred per second. Bandwidth is always a limited usable resource as far as radio frequencies are concerned, resulting in the need for regulation or a market to ration its use. Other media can overcome bandwidth limitations by increasing physical resources, such as laying additional cables or opting for higher bandwidth media like fiber-optic cable.

Baseband   audio or video signals before any modulation; generally extending to low frequencies. Audio baseband is 20 to 20,000 Hz; video is 20 Hz to 6 MHz.

Baud   a unit of speed for data communications. This is the number of times per second a signal is altered. Baud is usually equivalent to bit rate, but in some systems coding techniques make it possible to send more than one bit at a time, making the bit rate higher than the baud rate. See bit rate.

Baud rate   the speed of transmission between two devices

BBS   bulletin board systems on the Internet that provide electronic bulletin board and conferencing services

Bird   communications satellite

Bit   a digit of a binary system with only two possible values, 0 or 1

Bits per second   see BPS

Bit rate   bits per second

Blanking   in a video signal, the time period when picture information is shut off. Analog blanking is a voltage level that is at or below black picture level and acts as a signal to turn off the scanning beam. Synchronizing pulses, which control invisible retrace of scanning, are active during the blanking period.

Blocking   term used to describe a visual effect in which the TV image appears to be made up of blocks (typically 8 × 8 or 16 × 16 pixels). This unwanted defect is caused by a compression codec being unable to adequately process the data given to it, and thus the coding blocks become visible.

Bounds   hardware bounds describe the maximum pixel size of the on-screen display, regardless of the area covered by a region. Bounds are provided so that the application can always determine the size of the on-screen display and, therefore, position graphics relative to the on-screen display and to the video plane. This lets an application provide a unified look and feel, independent of the bounds across multiple platforms. On a PAL TV, the hardware bounds are always (0, 0) and (720, 576), and on an NTSC TV, the hardware bounds are always (0, 0) and (720, 480). In half-resolution mode, the bounds are (0, 0) and (720, 288), and (0, 0) and (720, 240), respectively.

bps (bits per second) a measurement of how fast digital data is moved from one place to another

Break the time available for advertising or purchase between two broadcast programs or between segments of a single program

Broadband a system of very large bandwidth. In ITV, the best example is cable TV, which can deliver multiple channels and services.

Broadband communications systems any technology capable of delivering multiple channels and services; often synonymous for cable TV.

Broadcast the practice of distributing a signal from one source to many receivers

Broadcast signal the signal that delivers radio/TV/ITV to a receiver set-top box. In a digital delivery, the multiplexed bit stream is transmitted across a network to all subscriber locations. Set-top boxes capable of descrambling the signal can then display the signal content to the attached TV screen.

Browser a client program (software) on a computer, set-top box or other device used to present various kinds of data resources; currently applies to looking at Web sources but expanding in use to cover the interface program for local files and other sources. May become a more generic term to cover certain ITV interfaces. The most common Web browsers for desktop computers are Netscape Navigator from Netscape Corporation and Microsoft Internet Explorer. These products are given away free. Browsers can be enhanced by third-party plug-ins. See also electronic program guide.

Browser caching to speed Web surfing, browsers store recently used pages on a user's disk. If a site is revisited, browsers display pages from the disk instead of requesting them from the server. Another type of cache is used on videotext systems—a model for ITV. This system downloads the most popular pages every time the user turns on, or in anticipation of the most likely pages the user is likely to need next. This makes surfing less frustrating.

Byte a group of 8 adjacent binary bits

C-Band signal frequency range (3.7–4.2 GHz)

CA (conditional access) an encryption/decryption management scheme where the broadcaster controls access to services, especially pay TV; ensures authenticity and security in purchase transactions

Cache stores received information in local memory, where a browser can get to it fast

Cable modem a modem that connects a set-top box or computer to a cable TV network and provides a computer with a higher information rate (higher bandwidth) than a modem connected to ordinary telephone lines. A cable modem taps into one of the cable TV channels dedicated to data services. A group of users share the same bandwidth available on the channel that gives them the same facility as a fast office network (10 Mbps). Computers accessing the Internet through cable modems are able to download graphics and video rapidly, without the expense of renting a fixed link such as ISDN or T1.

Cable television multichannel TV distribution via coaxial cables or fiber-optic cable. See CATV.

Camera electronic apparatus that converts the optical picture to an electrical signal (video)

CATV (cable television) originally community antenna TV, now used synonymously with cable TV

Cause effect chain in narrative structure, the way one event leads to another and is the result of a previous event. In certain ITV productions, this may become a complex network of possible causes and effects.

CBL computer-based learning

CBT computer-based training in which the computer becomes a tutor for asynchronous training that adjusts to each student's pace. While learning is predominantly for imparting knowledge, the emphasis of training is on acquisition of skill.

CC (closed caption) TV with simultaneous text broadcast, namely teletext. Also widely used to provide text for the hearing-impaired.

CCITT (Committee Consultatif International Telegraphique et Telephonique) committee of the ITU, now renamed ITU-R

CD (compact disc) a 12-cm laser-encoded optical disc for distribution of prerecorded music, originally developed by Sony and Philips. Since the original music coding standards, other standards have been developed for data storage, such as CD-ROM and CD-I (a Philips interactive CD standard). Originally a read-only technology was available. Now writeable versions are available.

CD-I (CD interactive) compact disc (developed by Philips, the electronics conglomerate headquartered in the Netherlands) that played back visual as well as audio CD entertainment and learning materials in a hypermedia format on a TV set. CD-I machines are no longer being sold and are of interest only as an example of an early ITV-like system.

CDMA/TDMA code division multiple access and time division multiple access dual-mode cellular telephones that aid in the receiving of fax and computer network data on computers and personal digital assistants.

CD-ROM (compact disc read-only memory) a 13-cm laser-encoded optical memory storage medium used to store data that computers can read. Contains 650 MB of data. Read-only memory means that the CD can only be read: unlike a computer floppy disk, the user cannot "write to a CD."

CD-R or CD-RW term used for machines (drives) that will record CD laser discs that will read on standard CD-ROM drives. CD encoding depicts the recording (burning or transfer of files) to a CD, whereas CD decoding depicts the reading of those files.

Central office a telecommunications facility where calls are switched

CGI   (common gateway interface) a specification for how an HTTP server should communicate with local programs on the server (called CGI programs), which are invoked by the HTML code in a Web page

Chat   instant text communications over an electronic network between users, either anonymously or with known correspondents. Today, chat is one of the most widely used services. More advanced forms of chat systems use 2D and 3D interfaces with avatars, graphical representations of each user that move in a graphical world. Chat can be used for live ITV channels. It could be an important feature to be built into ITV set-top boxes but demands Java interfaces and a keyboard.

Channel   a signal path of specified bandwidth and other properties

Channel capacity   maximum number of channels that a system can carry simultaneously

Channel surfing   changing channels on a TV to scan programs

Chrominance   the part of a TV or video signal that carries the color information; there are two elements of chrominance information

CHUT   cable households using television

Circulation   number of households with TV sets in the target area

Click stream   the stream of visitors to an Internet Web site

Click through   the action of using a hyperlink to jump from one part of the Web to another. Advertisers are concerned about click through, as it represents success in attracting visitors to their Web site.

Client   a computer that has access to services over a computer network. The computer providing the services is a server.

Client program   a software program used to contact and obtain data from a server software program on another computer over a network. A client program is designed to work with one or more specific kinds of server programs. A Web browser is an example of a client program.

Client-server a corporate computing trend that is gradually replacing the old way of conducting business, which was large mainframe computers connected to terminals. In the new arrangement, company software applications run on a mid-range computer (the server) that is connected over networks to PCs (clients).

Client-server architecture an information-passing scheme that works as follows: a client program, such as Mosaic, sends a request to a server. The server takes the request, disconnects from the client, and processes the request. When the request is processed, the server reconnects to the client program and the information is transferred to the client.

Closure a sale resulting from following up on an inquiry from direct mail advertising; could apply to sales from targeted advertising on an ITV service

CMC computer-mediated communication

Co-ax abbreviation for coaxial cable

Coaxial cable high-bandwidth copper cable used as transmission cable in TV applications. It is a concentric transmission line composed of a conductor centered inside a metallic tube or shield, separated by a dielectric material, and usually covered by an insulating jacket. A coaxial cable is capable of carrying many TV or radio signals simultaneously.

Code a historically and culturally based set of conventions in media

Codec (compression-decompression) a device that converts analog or digital video and audio signals into a digital format for transmission over telecommunications facilities and also converts received signals back into analog or digital format. This is key technology for making better use of limited bandwidth and storage capacity. Much information, especially video and audio, needs compressing (making the data rate smaller) to make it manageable.

Collaboration a joint effort that network technology has facilitated with e-mail, FTP, and more advanced means of sharing ideas, documents, and data. Writing has become more of a collaborative effort since the dawn of the World Wide Web. The Web takes authors beyond the telephone by enabling them to speak to one another (audio), see one another (video conferencing), and visualize documents and data. In an ITV environment, collaboration is required between network operators, service providers, and content developers.

Commercial impressions   the total audience for all advertisements in a schedule

Compression   in digital systems, the process of data manipulation to reduce the amount of data needed to represent information, such as a text file, a still image, motion video, or audio. See lossless compression, lossy compression.

Compression ratio   the ratio between the input and output data rates for a real-time compression module

Component video   a video signal in which the luminance and chrominance are sent as separate components, either as RBG (red, blue, green) or as luminance and color difference signals (e.g.. Y, Cb & Cr)

Composite video   a signal in which the luminance and chrominance are combined with synchronization signals using an encoding standard (e.g., PAL, NTSC, SECAM)

Concatenation   compression algorithms are designed to economize on bandwidth by eliminating parts of an image in a way that is least noticeable to the human eye. However, if a series of compression and decompression stages are cascaded (concatenated), then artifacts will become noticeable, particularly if each stage uses a different encoding method.

Connection orientated   circuit-like telephone, such as ATM, that uses switching techniques

Connectionless service   where packets are sent independently with no fixed connection, such as TCP/IP

Content   the actual information sent over a transmission system

Context definition   in order to help a user predict how a system will respond, some indication of how previous user actions or computer processing will affect the results of current actions.

Converter term for a set-top box; enables a conventional TV to receive multiple channels from a CATV feed

Cookie a piece of information sent by a Web server to a Web browser that is saved and sent back to the server whenever the browser makes additional requests (visits the Web site again) from the server. It may contain information such as registration, passwords, and user preferences. They are used to provide customized information to users, allow access to Web sites without the need for re-registration, and track visitors to, and usage of, Web sites.

Copyright the exclusive legal rights to perform or sell a movie, song, book, script, photograph, or any intellectual property. The copyright holder is paid a fee or royalty when the material is used. A multimedia production may include material covered by many different copyright regimes and can be a complex and expensive part of production. If there is no copyright, the material may be used for free and is said to be in the public domain.

CORBA (common object request broker architecture) designed to distribute objects or assembly of applications from discrete, self-contained components. The architectures of CORBA provide mechanisms for transparent invocation and accessing of remote distributed objects.

CPM (cost per thousand) a price comparison that shows the relative cost of various advertising media, from cost of advertising to a thousand households or individuals. For example, a Web site that charges $30,000 per banner and guarantees 600,000 impressions has a CPM of $50 ($30,000 divided by 600).

CPR (cost per rating point) the cost of advertising to 1% of the target audience

CPU a programmable unit in a computer that performs processing in the system; in a PC, the CPU is a microprocessor

CRM customer relationship management system based on information technology. Central to CRM are databases with sufficient details so that management, salespeople, people providing service, and perhaps the customer directly can access information, match customer needs with product plans and offerings, remind customers of service requirements, know what other products a customer had purchased, etc.

Cross-platform   the ability of a software package or an electronic book to run in more than one operating system

Cultural studies   a critical approach to understanding media. The discipline can provide many insights into understanding media and media use relevant to the development of ITV.

Cumulative audience   the unduplicated audience of a media service over more than one time period; also covers the number of different households or people reached by an advertising schedule

Cursor   a marker, such as an arrow, used to show the position of the user's operation on the screen. A device used in computer interfaces that can be used in ITV interfaces.

Cyber mall   term commonly used to describe an electronic site shared by a number of commercial interests

Cyberspace   term originally coined by author William Gibson in his novel *Neuromancer*, used now to describe the electronic space behind the screen. Alternatively, cyberspace is all of the information available through computer networks. The space metaphor is extremely interesting, and many uses of multimedia technology are direct analogies with real space, such as navigation systems, virtual chat rooms, and shared workspace. Interfaces for computer systems often use the space navigation metaphor.

DAB   digital audio broadcasting. See digital TV.

DAC   (digital-to-analog converter) usually a chip used in CD players and set-top boxes to change the signal to an analog form for reproduction of audio or video signals through speakers or TV

DAT   (digital audio tape) used to record CD files onto a cheap backup and storage medium. DAT tapes are contained in small cartridges that are the cheapest means of storing vast amounts of data. For example, a cartridge smaller than the palm of an adult hand can hold two or more gigabyes of data.

Database   a repository for data. A database is a computer file or system of data organized in records and fields for fast retrieval and easy updating.

Database technology is central to many new media technology developments. Multimedia databases, including those for pictures, text, sound, and video, are as complex as the transmission systems for delivery.

Data compression   see compression

Data conferencing   see whiteboarding

Data mining   manipulating data in a database to extract valuable information. Can help identify trends and relationships. Could be an essential part of the infrastructure that makes ITV a commercial proposition of service suppliers and advertisers.

Data protection   (1) data protection law: legal protection of personal or commercial data; (2) technology: to protect against data loss due to computer failure, user errors, unlawful access, and deliberate damage

DAVIC   (Digital Audio-Visual Council; http://www.davic.org) a nonprofit organization that develops and agrees to specifications of open interfaces and protocols that maximize interoperability across countries and applications/services to favor the success of emerging digital audio/visual applications and services. DAVIC 1.0 covers systems and applications such as TV distribution, VOD, and NVOD. DAVIC is open to any firm or corporation and includes content, service, and manufacturing companies.

DBS   (direct broadcast satellite) a system in which signals are transmitted directly from a satellite to a home rooftop-receiving dish

Decoder   in digital compression, the decoder is an electronic device used to convert an encrypted or compressed TV signal in a viewable picture. Sometimes used for set-top box. In broadcast teletext, device that displays closed captions hidden in the vertical blanking interval.

Definition   resolution, image quality, the ability of the system to display detail

Delivery mechanism   the medium through which the consumer can connect to a service. Delivery systems include broadband cable, satellite, ASDL.

Demographics   the aggregate characteristics of an audience or market, usually expressed in terms of age, gender, income, spending patterns, race, etc.

Descrambler   electronic circuit used to restore a scrambled video signal to its standard form. Part of the set-top box that decodes the scrambled video signal on a protected broadcast.

Dial-up connection   the most popular form of Internet connection for the home user, this is a connection from a PC to a host computer over standard telephone lines

Digital   method of storing, converting, and transmitting data in binary numbers

Digital compression   a digital TV signal is processed so that it requires a smaller portion of spectrum for its transmission. It could allow many channels to be carried in the capacity currently needed for one signal. See compression.

Digital media   refers to video and audio after having been digitized into 1s and 0s

Digital set-top box   a device for linking standard TV sets to the digital TV/broadband ITV network. Digital set-top boxes rely on data streams encoded using the MPEG standard. Digital set-top boxes are capable of two-way data streams, allowing consumers to access services such as electronic shopping and VOD.

Digital TV   television signals delivered in a digital form. The advantages of digital transmission are increased quality and reduced bandwidth. In addition, digital broadcasts allow the development of ITV services. Digital coding technology means that the picture never gets fuzzy, albeit it may disappear altogether when interference gets too bad.

Digital video disc or digital versatile disc   See DVD

Digital video effects   (DVE) a digital processing device that manipulates video signals in real time in order to create TV effects such as image shrink, warp, page turn, rotating box, etc.

Digitization   art or science of converting analog video, images, or audio into digital format. See ADC.

Digitize   to turn information from an analog format into a digital format to create an electronic copy of a picture or sound. See ADC.

Direct advertising   advertising under control of the advertiser, for example, telesales, direct mail, and e-mail or free sampling. ITV systems could make this a very important and cheap advertising method.

Direct mail advertising   advertising sent by post to an individual, household, or institution. Also advertising in other media that solicits orders directly through the post.

Direct marketing   sales made directly to the customer, rather than through intermediaries or intervening channels, includes telemarketing, direct mail, direct advertising. Some ITV systems may permit direct marketing (such as via the Internet), others may always interpose an intermediary.

Directory advertising   Yellow Pages advertising that appears in a buying guide or directory. One of the strengths of ITV is to make directories very accessible to consumers.

Dish   antenna using a parabolic reflector to collect microwave signals

Display structures   information elements that are consistent in appearance across applications. This may include functions to provide reference to the user's location, remind the user what options are available, and provide a visible boundary for user actions, such as menus and control buttons.

Display tailoring   designing displays or interfaces to suit a specific task or user, rather than providing a general usage display.

Distortion   any undesirable change in a waveform or signal

Distributed network computing   where a network computer can perform computing functions in another computer on the same network

DMA   (designated market area) an area based on those cities in which stations of the originating market account for a greater share of the viewing households than those from any other area

DNS (domain name server) refers to a database of Internet names and addresses that translates the names to the official Internet Protocol numbers and vice versa

Domain name unique name identifying an Internet site. Domain names always have 2 or more parts, separated by periods. The part on the left is the most specific, and the part on the right is the most general. It is necessary for Internet management that no domain names are the same. This requires a set of rules and final arbitration for disputes. See DNS.

Down link the segment of a satellite transmission from the satellite to an earth station or receiving antenna (dish)

Downstream signal path from cable head end or satellite to the subscriber. Generally, it has much higher bandwidth than upstream path. Conventional TV systems have no upstream path and many ITV systems assume that users will not require much upstream capability.

Drop cable piece of cable feeding into the subscriber's home

DRAM (dynamic random access memory) a chip used for temporary memory in the receiver

D-STB see digital set-top box

DSL (digital subscriber line) a technology for transmitting data over telephone lines, up to 50 times faster than present analog modem and ISDN alternatives. Telephone companies are hoping that DSL service will keep telephone lines competitive with cable modems and other competitive alternatives to current telephone transmission services. Telephone companies are considering two dominant DSL technologies: asymmetric digital subscriber line (ADSL, ASL) and high-rate digital subscriber line (HDSL). ADSL technology will deliver higher downstream speeds (6 Mbps) than upstream speeds (640 Kbps).

DSM-CC an MPEG standard for interfaces and control signals for set-top boxes, servers, and networks. The standard will provide for client-server operations such as service "navigation," video control, file, and database access.

DSS (digital satellite systems) for example, the DirecTV(R) system introduced by Hughes Electronics, Inc.

DTT (digital terrestrial television) digital TV broadcast from ground-based transmitters. See digital television and digital cable.

DTV See digital TV.

Dual cable two independent cable TV transmission systems operating side-by-side providing double the channel capacity of normal cable

Dumb terminal terminal that cannot store or manipulate data, only display it on a screen, such as a TV set or Minitel terminal

DVB (digital video broadcasting) standards developing forum of the European Broadcasting Union set up to define specification of baseline system for digital television broadcasting—DVB-S by satellite, DVB-C by cable, DVB-T by terrestrial mode

DVD (digital video disc) a new CD-based standard that provides the increased data storage capacity necessary to deliver feature-length films and programs at MPEG-2 quality on a single disc

Dynamic bandwidth allocation a transmission system that allocates a channel of the required bandwidth for each transmission

Earth station large ground-based satellite dish used for receiving and or transmitting to a communications satellite

E-business a term that is easily confused with e-commerce. E-business is an umbrella term that refers to any type of business transaction on the Internet. Some people may use the term in a more restricted context, such as a business-to-business (B2B) transaction as opposed to a business-to-consumer (B2C) transaction. For example, a B2B transaction might be a business firm's on-line banking transaction. E-commerce refers more to the B2C context where a firm sells goods on the Internet and makes collections via some payment scheme, such as on-line credit card transactions.

EDI abbreviation for electronic data interchange or electronic data invoicing. The EDI system allows linked computers to conduct business transactions such as ordering and invoicing over telecommunications networks.

Edit suite a nonlinear editing (NLE) computer or at least two videotape machines designed to function in a unit for editing purposes

Edit to insert, assemble, delete, or revise stored video or audio. Generally used to create a complete program from a number of individual shots or segments.

Edit decision list (EDL) a list of edit decisions accumulated in a video editor. The list typically includes the source, in-time, and out-time of each edit.

EFP (electronic field production) TV production activity outside the studio usually shot for postproduction

EFP camera a portable video camera, usually with a built-in recorder (then called a camcorder). An EFP camera may be of higher quality or offer more than a basic ENG camera.

Effective frequency and reach the level or range of audience exposure that provides what an advertiser considers to be the minimal effective level, and no more than this optimal level or range

Effects theory a communications theory that proposes that viewers are passive and that TV directly affects them

EIA Electronic Industries Association in the United States

EIRTT (European Industry Round Table on Information Technology) industry club to promote the development of digital cities—the development and use of network IT for the mass market, metropolitan life, and administration

Electronic mail see e-mail

Electronic program guide (EPG) an application that provides an on-screen listing of all programming and content that ITV service subscribers have available to them for browsing and selection

E-mail electronic messaging system, capable of sending and receiving most forms of information, pictures, and documents, via the Internet or ITV

Encoder   device that codes or scrambles information before transmission

Encryption   Cryptographic conversion of data into a secure format in order to prevent any but the intended recipient from reading that data. There are many types of data encryption, and they are the basis of network security.

ENG   (electronic news gathering) videotaping or digital capture and transmission of news events or actualities

ENG camera   (electronic news gathering camera) a camera with a built-in video recorder or used with a portable VCR. Most commonly used by TV news crews. See EFP camera.

EPG   see electronic program guide

Error management   interface feature to enable the detection and correction of transmission or user errors

Ethernet   a very common method of networking computers in a LAN. Ethernet can handle either 10 Mbps or 100 Mbps and can be used with almost any kind of computer.

Expository mode   method of TV where the facts are assertively or even aggressively selected, organized, and presented to the viewer in a direct address

Extranet   a term depicting networks on the Internet dedicated to business communications between a vendor and its suppliers, customers, or dealers. The term originated from network pioneer Robert Metcalfe. Using the common format of the World Wide Web, companies, their suppliers, customers or dealers exchange data electronically rather than sending paper-based information back and forth. It is viewed as an Internet alternative to electronic data interchange composed of dedicated lines and software rather than the Internet. Extranets are a lower cost alternative to EDI.

FCC   (Federal Communications Commission) body responsible for regulating telecommunications and broadcasting in the United States

FDDI   (fiber distributed data interface) a standard for transmitting data on optical fiber cables at a rate of around 100 Mbps (10 times as fast as Ethernet, about twice as fast as T3)

Fiber optics   light transmission through optical fibers for communications and signaling. Fiber-optic transmission is immune to all electromagnetic interference.

Fiber-optic cable   flexible glass fibers that carry communications signals on light waves produced by lasers. They provide high-bandwidth, high-quality transport for digital or analog signals. Fiber-optic cables are used as part of (wired) modern telecommunications systems, as well as modern cable TV systems. Because they can carry large amounts of information in two directions, they have been used to create the infrastructure for several ITV trials.

Field trials   an important part of service or product development is to test it with users in the field. This method has been a key part of the development of ITV services, where real customers in real-life situations (homes, offices, schools) test the systems. Field trials involve technical testing and social science research on the system, use, and content acceptance by consumers and business partners in the trial.

File transfer   the ability to transfer text, graphics, software, spreadsheets, audio files, and video files over vast distances on computer networks such as the Internet

Firewall   refers to security measures designed to protect a networked system from unauthorized or unwelcome access. The term is used in a more general sense than its original use as a "firewall machine." Originally, a firewall was a dedicated UNIX gateway machine with special security barriers. It protected other machines on Web connections and dial-in lines. The idea is to protect a LAN of more loosely administered machines hidden by the firewall from hackers, crackers, and phreakers. Incoming packets from the Web are met with firewall barriers of various types that trash packets not allowed behind the firewall machine.

Firewall is a software that provides security to a computer or local network by preventing unauthorized people from having access to information on a computer, especially where that information is on computers that are potentially publicly accessible via the Internet or WAN.

Flash memory   nonvolatile, erasable, and rewritable memory used to store and run programs in set-top boxes. Can also enable software upgrades for the set-top box.

Flow   a narrative of the sequence of programs, commercials, news breaks, and so on. The overall flow of TV is segmented into small parts, which often bear little logical connection to one another. Many ITV services will interrupt the flow or modify it in important ways. It is possible that the ability to break the flow may have important effects on viewer behavior. In the United States, this is often referred to as continuity.

FM   (a) frequency modulation; (b) floor manager

Focus groups   group sessions conducted with potential consumers to discuss their interest in an actual or potential service or product. Simple focus groups are often used in marketing and product development.

Frequency   the number of times an average audience member sees or hears an advertisement; the number of times an individual or household is exposed to an advertisement or campaign (frequency of exposure); the number of times an advertisement is run (frequency of insertion)

Fringe time   broadcast time periods preceding or following prime time. Television time between daytime and prime time is called early fringe and TV time immediately following prime time is called late fringe.

FSFMV   (full-screen full-motion video) full-motion video on a computer can be achieved relatively easily if the viewer is satisfied with a small (1/4 screen or even 1/16 screen) size for the clip. Achieving full-screen video is more difficult due to the larger amount of information that needs to be processed.

FSN   See full service networks

FTC   Federal Trade Commission in the United States

FTP   (File Transfer Protocol) an Internet technology that allows computer files to be transferred from one computer to another. See also HTTP.

FTTC   (fiber-to-the-curb) a transport architecture for providing interactive broadband services over optical fiber networks. Transport of the data streams from the central office to a neighborhood is over fiber. At the neighborhood, the signal is then carried over a short distance on standard twisted pair or coax lines to the individual homes. Because the distance is so short between the fiber's end and the subscriber's home, there is very little compromise on signal quality.

FTTH (fiber-to-the-home) same as FTTC, except the optical fiber network extends directly to the subscriber's home

Full duplex the channel allows transmissions from both ends to take place simultaneously, for example, telephone lines

Full motion video a video signal that plays at standard speed (24 frames per second or more) and not at the slower speeds that result in jerky motion

Full service networks proposed form of ITV services that offers what ITV companies consider to be a full range of services that includes home shopping and banking

Gateways connectors between two or more dissimilar networks that facilitate communication in such instances. Gateways have their own processors to perform both protocol and bandwidth conversions. Gateways between the Outernet and the Internet translate different protocols, such as e-mail protocols on different networks into Internet protocols.

Geostationary an orbit used by communications satellites. The geostationary (also called geosynchronous) orbit has an altitude of 35,785 km or 22,300 miles. At this altitude, the orbital speed matches the earth's daily rotational speed, so when the geostationary orbit is viewed from the earth, it appears to remain fixed at a relative point in the sky to the earth's surface.

General purpose interface (GPI) a parallel interconnection scheme that allows remote control of certain functions of a device. The general purpose interface bus (GPIB) was specifically designed to connect computers, peripherals, and laboratory instruments so that data and control information could pass between them. It is also known as IEEE-488, or HPIB, and is electrically equivalent to IEC-625 bus. It is defined completely in the IEEE standard 488.1-1987.

GHz gigahertz—a unit of frequency equal to 1×10^9 Hertz.

Gross audience the total number of households or people who are reached by an advertising schedule, including duplication

Gross exposures the total number of hits to the Web server resulting from a request for one Web page, including its graphic elements, sound files, or any other file called for in delivering the page for viewing through the browser. See hit.

Gross impressions the total number of audience impressions delivered by an advertising schedule

GRP (gross rating points) the total number of broadcast rating points delivered by an advertiser's TV schedule, usually in a one-week period. GRP is an indicator of the combined audience percentage reach and exposure frequency achieved by an advertising schedule.

GSM (global system for mobile communications) a variant of TDMA, GSM is the closest to a worldwide standard for cellular service. A single-frequency GSM cellular handset may work compatibly in Europe, Asia, India, and Africa, although not in the United States

GSO geosynchronous orbit. See geostationary.

GUI (graphical user interface) in computers or computer-like devices, a system for user-computer interaction that includes a video display and a pointing device such as a mouse. The user controls the system by pointing at objects (icons) on the screen and clicking a button.

HBI (horizontal blanking interval) unused gap in a TV signal that occurs before each line of a picture

HDSL a technology allowing two high-speed bidirectional channels for audio, video, data, and text transmission over T1/E1 lines

HDTV (high definition television) a TV signal with greater detail and fidelity than current TV systems use. HDTV provides a wide-screen picture with four times the visual resolution as current TVs, as well as CD-quality multichannel audio.

Head end electronic control center of a cable TV system. This is the site of the receiving antenna and the signal processing equipment essential to proper functioning of a cable system. Cable TV channels received from a local cable

TV company's satellite downlink are packaged together and transmitted to subscribers' homes. The telephony model equivalent is the exchange.

Head of household   the person within a family or household who is responsible for the major purchase decisions. In marketing, the male head and female head of household are sometimes considered separately.

Herz   (Hz) a unit of frequency, one cycle per second

HFC   (hybrid fiber coax) a digital transmission network composed of an optical fiber backbone connected to coaxial cable TV lines running to subscribers' homes. HFC is becoming a more popular solution for ITV networks as it takes advantage of existing cable TV coax connections.

High definition television   see HDTV

Hit   a very simple measure of traffic on a Web site. One hit represents the downloading of one file to another computer on the Internet.

Home page   the opening page of a Web site

Home shopping and banking   the ability for a TV viewer, telephone user, or computer user to undertake shopping or banking remotely, such as from the home. The implication is that the TV, computer, or other terminal can connect and exchange information with the bank or shop via the Internet. Home shopping has long existed on the Internet via PCs and TV, using the phone as the return path for orders. The main innovations for home shopping or banking is not the home terminal or connection but the business system that displays and markets the goods, handles transactions, and delivers the goods, as well as after-sales service.

Host   any computer on a network that is a repository for services available to other computers on the network. On the Internet, a host can provide Web, news group, and e-mail services. Network Wizards defines a host as a domain name that has an IP address record associated with it. The terms host and domain are used interchangeably. The definition of host is changing due to virtual hosting, where a single machine acts like multiple systems.

Hot link   see hyperlink

Hot spots   parts of a text, video, picture, or audio document designated by the author to be active areas that, when selected by the user, performs some action, such as retrieving and downloading another document. Hot spots are an important form of interactivity on many multimedia systems and ITVs. These can be used in many ways, for example, to bring up additional information on a product advertised; link to alternative information on a news story; or to engage in a different part of a fictional environment in an ITV production or video game.

HTML   (Hypertext Markup Language) the language used to tag various parts of a Web document so browsing software will know how to display that document's links, text, graphics, and attached media

HTML document   a document coded with a Hypertext Markup Language

HTML engine   a software platform for deploying numerous ITV applications, including Web browsers, e-mail, electronic programming guides, chat, on-screen billing, and on-line shopping

HTTP   (Hypertext Transfer Protocol) The Web's protocol that tells the server what to send to the client, so the client can view Web pages

Human factors   human issues in designing a system; generally used in respect of cognitive and individual issues, occasionally extending to social and economic factors

HUT   households using television

Hyperlink   text on a Web site that can be clicked on to go to another Web page or a different area of the same Web page. Hyperlinks are created in HTML.

Hypermedia   Hypertext with added features for audio and video. Hypermedia may also entail touchscreen or remote control capabilities such that users can navigate by touching the computer screen or remote control devices. Eventually, hypermedia will entail other senses such as smell. The key to hypermedia is random access that allows lightning-fast nonlinear navigation based on reader choice or other reader actions such as responses to questions. The term multimedia is not totally synonymous with hypermedia, because multimedia may not entail hypertext authoring. In an ITV environment, hypermedia can be used in many ways, namely to bring up additional information on an advertised product, to retrieve alternative information on a news story, or to engage in a different part of a fictional environment such as an ITV game or drama.

Hypertext   See hyperlink

Icon   graphical representation of an object (file, directory, picture, text field, etc.) as a tiny symbol that can be arranged with other icons and clicked on using a pointing device. See GUI.

IEEE 1394   (serial interconnection bus—iLink) a high-performance serial connection. A high-speed serial digital interface standard enabling data communications between the digital set-top box and DVD players or D VHS recorders. Transmission speed is scaleable from approximately 100 Mbps to 400 Mbps.

IETF   (Internet Technical Forum) a subgroup of the Internet Architecture Board that focuses on solving technical problems on the Internet

Impressions   see ad views

Information appliance   a term used to describe a domestic set-top box with broader information technology capabilities than a simple TV decoder

Information superhighway   a worldwide combination of fiber-optic cable and satellite receivers in a future time when homes and offices around the world will be linked by highways of electronic information that can be traversed interactively both to and from a connected user. Technologies are coming to a head and plans are being laid to create the digital information highways. In homes and offices, a single piece of interactive TV digital TV equipment (let's call it the PCTV computer or television superhighway terminal) will combine what are now TV sets, telephones, stereos, videotape players, videodisc players, CD players, and computers. The PCTV will be networked to hundreds of millions of servers ranging from the computer files of individuals to the systems of computer files that contain virtually all the movies ever made, all the contents of daily newspapers, all the television shows ever recorded, all the cataloged products and services available from vendors, all public documents of governments, all the contents of libraries, all instructional and training courses on anything known in the world, and so on to limits beyond our current imaginations.

Innofusion used to describe the way that innovation occurs as an integral part of the diffusion process. It emphasizes the importance of users in developing uses of new technologies and of reinnovation by users and suppliers. This is contrary to the traditional view of technological development, where innovation of technology precedes diffusion. Interactive television illustrates the innofusion process very well.

Interference undesirable signals that are the result of electromagnetic radiation. These undesirable signals degrade the quality of desired signals.

Interactive TV (ITV) a term that refers to the convergence of computing, communications, and entertainment on a TV. Examples of services that would be offered with ITV include shopping, e-mail, gaming, VOD, access to local community information, and electronic programming guides. More specifically, it is defined as a TV system that allows the viewer to send information to the broadcaster, as well as receive information from the broadcaster. By sending information, viewers can request information and programs, change the content of programs, and undertake on-line activities such as home shopping and home banking.

Internet global computer network based on nonhierarchical data interchange technology (TCP/IP). It is now a worldwide collection of independent, interconnected networks of computers, communicating through a common and simple language using telephone lines or other forms of network. Internet technology supports many different applications (i.e., e-mail, Web, news groups, live audio and video).

The Internet has become an important factor in the development of mass market ITV.

Internet 2 a consortium of more than 100 universities and other organizations collaborating to develop next-generation Internet technology. In addition to bandwidth issues, the consortium is dealing with such issues as audio and video integration, interactive distance learning, telemedicine, on-line research collaboration, and real-time simulation or modeling.

Internet audio and video the transport of audio and video such that users can hear sounds and watch video while reading text and graphics on the

Internet. It is no longer necessary to download these media files and install them on a local computer or server. These files can be played live on the Internet. Real-time delivery, or streaming, is now common where high-speed connections are available.

Internet messaging　technologies for sending messages across the Internet.

Internet service provider　(ISP) companies that sell Internet access, including software, that generally enables users to access the Internet through a local telephone call or DSL, which does not require dial-up. Users get their computers to telephone the ISP and they are then routed onto the main backbone of the Internet and, from there, to remote locations.

Interoperability　in a digital TV system, interoperability permits operation on a variety of media and between equipment from different manufacturers

Intranet　use of Internet technologies for private and internal communications by organizations

I/O　input/output

IP　the abbreviation for Internet Protocol, IP refers to the set of communications standards that control communications activity on the Internet

IP address　a computer's address on the Internet, expressed as a series of numbers

IPPV　(impulse pay per view) a function that allows a viewer to choose at the time of viewing to pay for a particular program. The user does not have to prepay or subscribe to the channel. IPPV can become an important revenue stream for ITV services and will allow independent ISPs such as TV production companies to derive revenues directly from the viewer. See pay per view.

IRC　(Internet relay chat) a worldwide network of people "talking" to each other in real time via the keyboards of their computers. Chat groups can be anonymous and banal or develop into regular "communities."

ISDN　(Integrated Services Digital Network) a digital network technology that moves up to 128 Kbps (ISDN 2) over a regular phone line. A standard

ISDN line carries 64 Kbs. ISDN offers other services, such as call waiting, that analog networks in some countries do not offer.

ISOD   interactive service on demand

ISP   see Internet service provider

ITC   (Independent Television Commission) U.K. television regulator

ITU   (International Telecommunications Union) an international organization within which governments and private sectors coordinate global telecommunications networks and services

ITV   see interactive TV

Java   a computer language developed by Sun Microsystems; designed for programs that can be downloaded over a network to run directly on any machine without the need for operating system compatibility. A Java program needs a Java "virtual machine" to run on, or JavaOS. The virtual machine links the program to the underlying operating system. Java is also intended to run on other computing devices such as ITV set-top boxes.

Java is based on object-orientated programming theories, which make code writing more efficient. On the Web, small Java programs, or applets, help get a user's profile. Java has been accepted in digital TV and ITV systems.

JavaScript   a simpler version of Java; a computer language that enables Web site designers to enhance Web sites with more interactive features

Java virtual machine   software that allows Java to run on different platforms

Jitter   instantaneous timing errors in the position of signal transitions. It can introduce errors or loss of synchronization in digital systems.

JPEG   (Joint Photographic Experts Group) a standard for the compression of still images and graphics, developed by the Joint Photographic Experts Group. See MPEG.

Jscript   a scripting language developed by Microsoft to enable Web authors to design interactive sites. Although it shares many of the features and structures of the full Java language, it was developed independently. Jscript can interact with HTML source code, enabling Web authors to spice up their sites with dynamic content.

Kbps   one Kbps is one thousand bits of information every second

Kiosk   a free-standing interactive computer or audio-video system used for public access or information delivery. Interactive kiosks are placed in public places, such as retail environments, railway stations, and corporate offices, for the public to find out information, conduct transactions, etc. This information may include classified advertising, tourism information, ticket sales, mail order services, etc. Generally, the user will touch the screen in appropriate places to access the information. See touchscreen.

KISS   (keep it simple, stupid) refers mainly to the authoring of electronic books and the development of software in which success often depends on keeping the learning and usage mindlessly simple

Knowledge management   a term that can have multiple meanings. In business information technology, knowledge management refers to an entire integrated system to accumulate, integrate, manipulate, and access data across multiple organizations, including such data as credit data, consumer profiles, market data, product development data, etc.

LAN   (local area network) a network of computers, generally in the same building, connected together with fixed cabling or through microwave links. A LAN can be as small as two computers linked together.

Laser   a device for generating a coherent source of light with a narrow beam and a narrow spectral bandwidth

Link   the ability for one Web page to call for another page or site to be displayed. Also called a hot link.

Listserv   an e-mail system where users subscribe to join in on group messages. A message sent to the listserv is sent to every subscriber's mail box. A listserv is similar to an e-mail bulletin board. However, users of bulletin boards do not receive the messages in their mail boxes without first going to the bulletin board to view a listing of messages.

Local area network see LAN

Local loop the local network link to subscriber's homes and business. The most expensive part of a cable TV network, it includes cables, curbside routers, local exchanges, or cable head end. Technologies include twisted pair, coax, FTTH, FTTC, and wireless networks.

Lossless compression a method of compressing information that does not involve the loss of data. One simple type of lossless compression, called *run length coding,* replaces repetitive data with tokens. For instance, a string of data ABBBCDDDDDDE would be coded A3BC5DE. Lossless compression is less effective than lossy compression, but results in perfect reproduction of data.

Lossy compression involves stripping out redundant data that will not significantly degrade the quality of the information when it is played back. For instance, a clip of video may consist of a person talking in front of a still background. If the background does not change from frame to frame, the computer only needs the data to indicate that no change has occurred. It does not need the data that makes up the background to be repeated for each frame. Lossy compression can sometimes create video with a compression ratio of 1:200. However, the greater the compression, the poorer the quality of reproduction.

Luminance the brightness component of a color TV signal

Mbps (megabits per second) one Mbps is one million bits of information every second

MDS multipoint distribution service

Media anything that carries or stores information, such as paper, newspapers, electronic discs, etc.

Megahertz (Mhz) unit of frequency equal to one million hertz (one million cycles per second)

Microprocessor an integrated circuit chip that processes, calculates, routes data and information, etc.

Middleware second-generation network computing applications extend data transfers from client computers back to the Web server and/or database server computers. Software for doing this is commonly termed middleware. Software mediates between an application program on a server and a network of client machines. Middleware manages the interaction between applications across the heterogeneous computing platforms of client computers.

MIPS (millions of instructions per second) a measure of the processor execution speed

Mode of representation manner in which a nonnarrative TV program depicts historical reality and addresses itself to the viewer about that version of reality. Modes include expository, interactive, observational, and reflexive. An ITV system can be used to offer the viewer a number of different representations simultaneously or engage viewers more intimately in an interactive mode.

Modem (modulator demodulator) a contraction of modulate-demodulate, most commonly used to refer to a device for turning digital information into analog information that can be passed down via ordinary telephone wires. Modems, therefore, interface computers to the Internet or another network.

Modulation altering the characteristics of a carrier signal to convey information; techniques include amplitude, frequency, phase, and pulse width

Monitor see TV monitor or audio monitor

MPEG (Motion Picture Experts Group) a committee of technical professionals from different industries dedicated to forming open standards for the transmission of digitized video and audio for computer and TV networks. MPEG-1 has been approved by the International Standards Organization (ISO) and has a transfer rate of 1.544 Mbps, compressed typically at 40:1 ratio. It is often used for low quality, such as that used to transmit video across the Internet or from a multimedia CD-ROM. MPEG-2 is generally compressed at 30:1, a similar quality to NTSC or PAL, but takes a higher data rate (6–8 Mbps) and is used for European digital broadcasting and DVD. MPEG-4 is used for video conferencing.

MSO   (multiple services operator) a cable TV company with more than one community network

Multimedia   refers to the delivery of information that combines different content formats (such as text, graphics, audio, still images, animation, motion video) or storage media (magnetic disk, optical disc, tape, RAM)

Multiplex   a technique for putting two or more signals into a single communications channel

Multitasking   execution of programs simultaneously on a single computer. In newer operating systems, two or more programs may be running in the background while the user is concentrating on another program running in the foreground. Limits on how many programs can be run at the same time depend more on hardware capacities, especially RAM amounts. Most operating systems now have multitasking capabilities.

muSIST   (multimedia user interface for interactive services and TV) ACTS project to develop and test user interfaces for home-based ITV. Includes lab work and testing on EU national hosts.

Narrowband   a data transmission system that has a small bandwidth. Traditional copper telephone cables are narrowband (offering 4 kHz analog).

Near video-on-demand   see NVOD

Netscape   now owned by AOL, the creators of the Netscape Navigator Web browser for desktop computers and an important player in developing the market for Web services, especially for developing secure transactions systems, a prerequisite for the serious development of Internet commerce and Internet shopping and banking. See browser.

Networks   linkages between computers allowing data and other digitized information to be transmitted between computers. Networks may be local, regional, national, or international.

Network computer   the name given to a computer that relies heavily on network facilities to store data and programs. Individual network computers would have less power and memory than conventional PCs and no storage disk, because smaller applications could be downloaded on demand and data would be stored on central or distributed servers. An architecture that appears to be in tune with moves to share data and move maintenance costs back to central corporate servers.

A network computer shares many features in common with the information appliance, a consumer electronics version for delivering ITV services.

News group a discussion group on the Internet devoted to talking about a particular topic; can be international or local, and the decision to carry (a copy on the local server) a news group is made by the local system administrator. However, a news group can always be accessed on another distance machine over the Internet. News groups are one of the important parts of the content on the Internet.

NII (national information infrastructure) projects to develop information superhighways for the economic and social benefit of the country. Although government-led, it relies on the development of multimedia-capable networks (especially broadband) by commercial companies.

NFS (network file system) a protocol suite developed and licensed by Sun Microsystems that allows different makes of computers running different operating systems to share files and disk storage

Node a workstation, file server, bridge, or other device that has an address on a network

Nonlinear editing (NLE) editing using a computer and associated disks, which allows fast access to any information stored on the disk. This is opposed to linear editing on videotape machines, which usually requires considerable time to fast forward or rewind to the required footage.

NTSC the composite color TV system with 525 lines, interlaced at 60 Hz as defined by the U.S. National Television Systems Committee; mainly used in the United States and Japan.

NVOD (near video-on-demand) proposed use of multichannel broadcasting facilities to try and satisfy a supposed consumer demand for flexibility in the viewing times of movies (films). Technically, the transmission of the same video simultaneously on several channels, but each starting a short time after the other (for example, 10 minutes), making it possible for the viewer to choose to start watching over. A term derived from VOD, NVOD is an ideal way to fill the hundreds of channels opened up by digital cable and satellite. Near movies-on-demand is currently being offered in cinemas, where large multiplexes can offer the same film (movie) on several screens, giving the audience the same benefits as its TV-based sister.

OB van a mobile TV production unit used for outside broadcasts. Usually equipped with control room, VTRs, and other recording equipment.

Off-line a situation when a computer is not connected to a network

On-line an expression denoting the fact that a computer or computer-like device is connected to a network. A PC that is connected to the Internet or a private service such as a banking or news service is said to be on-line.

Open a public standard in computer contexts that is the opposite of proprietary. Open refers to software and hardware made from published specifications that anyone can copy, giving customers a choice among multiple suppliers that compete on price and innovation.

On-TV™ a proprietary software system for bringing interactive services to the TV via a set-top box using TV-HTML, accessible from the Internet for broadcast. Produced by Viewcall, Inc., the system has been licensed to many hardware manufacturers to include in their set-top boxes.

Open cable a project aimed at obtaining a new generation of set-top boxes that are interoperable. These new devices will enable a new range of interactive services to be provided to cable customers.

Operating system the master control software system that serves as a foundation for applications software in computers or set-top boxes.

OEM original equipment manufacturer

Packet a discrete unit of data bits transmitted over a network

Page Web sites are referred to as collections of electronic pages. Each Web page is a document formatted in HTML that contains text, images, or media objects such as RealAudio player files, QuickTime videos, or Java applets. The home page is typically a visitor's first point of entry and features a site index. Pages can be static or dynamically generated.

Paging   a method of viewing and navigating data in which a user is presented with a metaphor that conceives of data as being grouped into easy-to-handle pages and moves through it by discrete steps

Page views   number of times a user requests a page that may contain a particular ad. Indicative of the number of times an ad was potentially seen, or "gross impressions." Page views may overstate ad impressions if users choose to turn off graphics (often done to speed browsing).

PAL   (phase alternating line) the color TV system with 625 lines and 50-Hz interlaced scanning; used in Europe, Australia, and many other countries throughout the world

PANS   pretty amazing new stuff

Parental lock   a feature of varying sophistication in receivers or PCs to prevent access by children to certain programs or services

Password   a secret word used to identify a user

Pay-per-view   (PPV) a system of offering TV broadcasts such that viewers pay for the programs they watch rather than subscribing to the whole output of a channel or broadcaster. PPV is often used to broadcast one-off events such as boxing matches. Currently, PPV systems rely on repayment by buying an access code on a card and have mixed success. One of the functions enabled by ITV is impulse pay-per-view, when viewers decide to pay at the moment of viewing, bringing consumption to the point of transaction in time and space. See IPPV.

PCM   (pulse code modulation) in digital video and audio systems, the representation of analog signal by its direct digitized sample values

PCMCIA   (Personal Computer Memory Card International Association) an interface standard for connecting peripherals such as modems, or any other multipin connection, to small computers such as laptops and handhelds

PCS   (personal communications service) refers to the three predominant digital cellular technologies operating in the 1.9-GHz band—CDMA, GSM and TDMA, all of which can allow data to be sent over cellular networks

PDA (personal digital assistant) a pocket-sized device for recording typed or handwritten messages that can later be ported to computers

Peripherals units or modules external to the basic unit in computers and computer-like devices, such as set-top boxes

Phase refers to the angular relationship between two synchronous signals and is expressed in degrees

PIP (picture-in-picture) the simultaneous display of two TV images on the same screen, including images from two separate TV tuners or a TV tuner and VCR tape deck

Pixel a point in an image. Each pixel has separately definable brightness and colors. An image is a two-dimensional array of pixels.

Port physical interface used to transfer data in and out of anything, such as a set-top box

Portal a Web site that offers a broad array of resources and services, such as e-mail, discussion forums, search engines, and on-line shopping malls. The first Web portals were on-line service providers, but now most of the traditional search sites have transformed themselves into Web portals to attract and retain a larger audience.

Postproduction in audio and video program creation, the process of assembling materials from production into a complete program. See production.

POP (1) (point of presence) a central facility or hub set up by an Internet service provider for subscriber access; (2) (picture-on-picture) entails widescreen viewing of up to three TV images simultaneously on 16:9 wide-screen TV

POTS plain old telephone systems

PPV see pay-per-view, IPPV

Production in audio and video program creation, the process of collecting (recording) material for use in the program

Protocol any formal description of message formats and the rules two computers must follow to exchange those messages

Proxy server a server that acts as an intermediary between a Web client and that Web server. A proxy server serves several purposes. It holds the most commonly and recently used content from the Web for users in order to provide quicker access. It filters Web content. It converts Web pages to match the capabilities of the receiving client or device. Another feature of these is firewall capability.

PVR personal video recorder

QAM (quadrature amplitude modulation) a modulation technique (for transmission of digital signals) that is often used by modems, including those that transmit digital TV

QPSK (quaternary phase shift keying) another modulation technique, similar to QAM, widely used for modems. Some ITV set-top boxes use QPSK for two-way signaling and messaging.

Radio frequency (RF) frequencies in the electromagnetic spectrum that are used for radio communications

RealAudio a commercial software program that plays audio on demand (see VOD) over the Internet. RealAudio and other technologies are used by many radio stations to broadcast live on the Internet or transmit long recorded tracks that would be too laborious to download at one go before listening. It is already possible to listen to radio stations from all over the world on a PC, with reasonable quality.

Real time in true or actual time, i.e., normal speed. A computer video file compressed, transmitted, and decompressed in real time is in effect being broadcast live.

Receiver a unit that takes signals from the antenna and tunes and converts them for TV display

Rec. 601 video sampling standard that includes the definition for 4:2:2 component digital video. Previously called CCIR Rec.601, it is now officially known as ITU-R Rec. BT.601.

Repeater signals sent over a transmission line get weaker as energy is lost during transmission. Signals also get distorted over distance. Repeaters are used to amplify and reconstruct the signal to enable it to be sent a greater distance. The advantage of a digital system over analog is that the signal is completely reconstructed by a repeater, removing noise and interference from other signals on the same line.

Resolution a measure of the ability of a TV system to reproduce fine detail

Response time the time between the user submitting a command to a system and the system producing the desired outcome. The user's expectations of response time and patience with the system can have considerable effect on usability.

Return path route through which the user of a set-top box can send data back to the broadcaster, such as via a modem, connected to a telephone line

RISC (reduced instruction set computer) a type of computer chip that uses a simplified set of instructions that can be combined to achieve more complex actions. RISC computers may be less expensive than conventional computers for the same processing speed.

RF modulator an electronic device that accepts a baseband signal, which modulates a carrier frequency resulting in the formation of a higher frequency signal that can be used for broadcasting purposes

RF tuning VHF/UHF tuning for TVs to select the desired channel

RGB a type of component video signal divided into separate (red, blue, and green) color channels

RS232 serial port connection suitable for computer links from PCs to modems and other peripherals; can transfer data over longer distances than parallel ports

Sampling rate the frequency with which samples are taken in digitizing operations. Sampling rates for audio range from 11 to 48 kHz, and for video from 13.5 to 75 MHz.

Satellite   for TV, device located in geostationary orbit above the earth, which receives transmissions from separate points and retransmits them to cable systems, DBS, and others over a wide area

Satellite news gathering   (SNG) video or audio news gathering using a mobile or transportable unit that uplinks a signal from a remote location to a satellite.

Scalability   relates to the ability of a system to grow

Scrambling   a signal security technique for rendering a TV picture unviewable while permitting full restoration with a properly authorized decoder or descrambler

SDI   (serial digital interface) see serial digital

SECAM   (SEquential Couleur Avec Memoire) color TV system used in France, Russia, etc.

Search engine   a piece of software that looks for strings of characters in a store of information and then identifies them for the user. Search engines are often used on the Internet to find documents containing certain words. Search engines are crucial to making the Web useful. Could become essential to ITV, if an open system develops.

Security   deals with protection against error and fraud. In computing and networking, this includes firewall protections, encryption, passwords, or other techniques to secure messages that contain sensitive or private data such as credit card numbers.

Secure socket layer   (SSL) a Netscape-developed protocol that operates with Internet systems to ensure security of transactions in services such as electronic commerce and home banking. Netscape developed SSL using encryption technology from RSA, Inc. It involves implementation of a software layer between the communications layer (TCP/IP) and applications layer (HTTP/FTP, etc.).

Serial digital   refers to digital information that is transmitted through a single wire. See SMPTE 259-M.

Server   a computer, or software on a computer, that makes services or information available to other computers on a network

Set-top box   (STB) a consumer electronics device that serves as a gateway between the TV and network. Any of several different electronic devices that may by used in a customer's home to enable services to be on that customer's TV set. If the set-top device is for extended tuning of channels only, it is called a converter. If it restores scrambled or otherwise protected signals, it is called a descrambler.

In ITV networks, the STB receives encoded (or compressed) digital signals from the network and decodes (or decompresses) the signals and converts them into analog signals for display on a TV. The STB also receives commands from the user (usually via infrared remote control) and transmits such commands back to the network.

STB proposed for ITV systems may combine many functions, such as TV receiver, games machine, modem, Web browser, EPGs, CD-ROM, or DVD player.

Signal-to-noise ratio   in signal systems, the ratio of the desired signal to spurious noise or interference produced in the system

Signal   the carrier of information through a telecommunications system

Smart card   a credit card with an embedded microchip that contains extensive information. Smart cards are currently used for telephone cards, health cards, pay TV, banking, GSM communications, and other cellular/satellite telephones. Smart cards can hold encrypted secure data transferred in from a PC.

SMPTE 259-M   Society of Motion Picture and Television Engineers standard for a serial digital interface for 525–60 or 625–50 digital TV equipment operating with 4:2:2 component digital signals at 270 Mbps

Social shaping of technology   a theory that professes that technologies are not developed as part of an inner logic of scientific progress. The development is rather the outcome of a process of negotiation between social actors attempting to shape the technology and its uses to their own ends and according to their own knowledge and priority, in the context of specific economic, social, and political environments. Extremely useful tool for understanding the emergence of new technologies and markets in periods of high uncertainty, when uses and users (market and demand), regulations, and industry are in flux.

SONET (Synchronous Optical NETwork) a family of fiber-optic transmission at rates from 5184 Mbps-1322 Gbps, created to provide the flexibility needed to transport many digital signals with different capacities and to provide a standard for manufacturers. Bellcore initiated SONET in order to enable multivendor interworking. The other goals were to be cost-effective for existing services on an end-to-end basis and to create an infrastructure to support new broadband services.

SSL see secure socket layer

SSM super slow motion

STB see set-top box

Storyboard a graphical representation of an interactive program. The same as a video storyboard but with the addition of a line representing hyperlinks between parts of the program. Used in authoring.

Subscriber customer paying a monthly fee to system operators for the capability of receiving a diversity of programs and services

Surfing browsing TV channels or the Web

Synchronous a method of communication using a time interval to distinguish between transmitted blocks of data

T1 a digital transmission link (network connection) with a capacity of 1.544 Mbps over two pairs of normal twisted wires. T1 can normally handle 24 voice conversations, each one digitized at 64 Kbps. ADSL and HDSL technologies are implemented on T1 transmission links.

T3 an even higher speed (45 Mbps) Internet connection

Tags formatting codes used in HTML documents. Tags indicate how parts of a document will appear when displayed by browsing software.

Target group the people at whom an advertising campaign is directed; individuals with similar characteristics likely to buy a product or service

Target market the geographic area to which an ad campaign is directed or a product sold

TCP (Transmission Control Protocol) works with IP to ensure that data packets travel safely on the Internet

TCP/IP (Transmission Control Protocol/Internet Protocol) an Internet transmission protocol that is the basis of the Internet. This is a standard for routing and data transfer around the world.

TDMA (time division multiple access) a method of dividing a single channel into a number of time slots and assigning each user a distinct time slot within a given channel. This lets more users (usually three) access a channel at one time without interference. TDMA is one of the standard digital cellular technologies, along with CDMA. GSM is a variant.

Teleconference telephone communications in which more than two people are simultaneously connected so they can exchange verbal comments as if they were in the same room having a face-to-face conference. A teleconference need not have visual communications in addition to audio communications, but modern technology now makes it possible to see conference members on monitor screens or TV screens.

Telemarketing selling by use of telephones, by initiating the calls (to a predefined target group or market), or receiving inquiries and orders. ITV opens up many prospects for telemarketing, both as a system for gathering marketing data, as well as for video conferencing with potential customers. ITV telemarketing is currently restricted to in-store terminal pilots.

Teletext a closed captioning or broadcast videotext standard used in the United Kingdom and parts of Europe. Teletext systems in the United Kingdom are very successful, simple versions of ITV, supplying news, weather, sports, games, ads, viewers' feedback, and other programs. Viewers access pages broadcast in a carousel fashion. Now teletext is being used for ITV services with feedback via the telephone. Many TVs come with teletext built in and the services are free at point of use.

Timecode   time-of-day or relative time information is recorded on professional media formats and used to edit and identify scenes

Time period rating   (TPR) the advertising rating for a particular broadcast time period, regardless of the program broadcast during that slot. Valuable because it recognizes that many people have similar time-based viewing habits based around other activities in their everyday life. With ITV systems, this time period could be estimated for individual viewers rather than extrapolated for an entire target group or population from a survey. VOD could change the value as different people watch different programs at the same time, according to their wishes rather than the broadcaster.

Touchscreen   a computer monitor's screen that is sensitive to touch. They are used in place of keyboards as an easy way of presenting information to people, especially those who are unfamiliar with computer keyboards or in situations where a keyboard would not be sufficiently robust. See kiosk.

Transport stream   MPEG-2 multiplex with short fixed-length packets carrying many programs intended for broadcast over potentially error-prone media, such as a satellite

Transmission   the feeding of a signal to a remote location

Transponder   a receiver and transmitter in a satellite that relays the signal from uplink to downlink

Trunking   transporting a number of signals from one point (an antenna site, for instance) to another point (such as a head end), usually without serving customers directly. Trunking can be accomplished by using coaxial cable, fiber optics or microwave radio.

TV-HTML   (Television Hypertext Markup Language) proposals for modifying the Web's HTML standard to make Web pages display better on TV screens, which typically have lower resolution than computer screens

TV monitor   TV display screen used to view a TV signal in its unmodulated baseband state. A TV monitor has no tuner and is not capable of receiving off-air or CATV signals without additional equipment

TV receiver TV display screen able to tune into a modulated or off-air TV signal

Twisted pair ordinary domestic copper telephone cables

Two-wire circuit telephone line or pair of wires to carry an audio circuit

UHF (ultra-high frequency) the spectrum extending from 300–3,000 MHz

Uplink signal path from the earth to the satellite

Unique resource locator see URL

Unique user the number of individuals who visit a site within a specific time period. To identify unique users, Web sites rely on some form of user registration or identification system. See cookies.

UNIX a computer operating system (the basic software running on a computer, underneath things like databases and word processors). UNIX is designed to be used by many people at once (multiuser) and has TCP/IP built-in. UNIX is the most prevalent operating system for Internet servers and ITV systems.

URL (unique resource locator) a Web site's address on the World Wide Web. Using this address, a user's Web browser is able to connect to a remote computer or Web server. URLs have a minimum of two parts but often contain three or four. A typical URL is www.air.net

User interface the layout of a computer screen or ITV screen. The design of interfaces to be used by people who are unfamiliar with computers is an important issue in making multimedia programs and ITV more popular. Work on voice recognition, which will allow users to talk to their computers, is progressing. The ultimate interface is a brainwave interface where people can control computers by "thinking at them."

User response time the speed with which a user can enter commands or control a system. It is not connected to the system's ability to respond or process the commands. See response time.

V modem   speeds governed by bps. V22: 1,200 bps; V22 bis: 2,400 bps; V23: 1,200 bps one way, but only 75 bps back to exchange; V32 bis: 14,400 bps; V34: 28,800 bps.

VALIDATE   (Verification and Launch of Integrated Digital Advanced Television in Europe) EC program lead by BBC R&D with 20 participants, including DT, Bosch, Thompson Brandt, Philips, RIA, and Rétévision

Valid hits   a further refinement of hits, valid hits are hits that deliver all of the information to a user. Excludes hits such as redirects, error messages, and computer-generated hits.

Vaporware   products and technologies that are announced but do not, and may never, exist

VBI   (vertical blanking interval) the blank lines of the TV broadcast between one frame and the next that can be used to carry extra information. Used for closed captioning and teletext. Increasingly, ITV systems (such as two-way TV in the United Kingdom) and Internet broadcasting systems (such as Intel's Intercast system) make use of VBIs.

VDSL   (very high bit rate digital subscriber line) a technology allowing very high speed bidirectional transmission of video and audio over the telephone network

VDT   (video dial tone) a term derived from telephone dial tone, to refer to on-demand transmission of video data by telephone companies between homes and businesses in a similar way to current data and telephony services. More of an open approach to multimedia service delivery than VOD, which is the term for a centralized media model of transmitting interactive video services. Could be a service provided for ITV, VOD, video conferencing, and transmission of video from the home.

Vertical Interval Time Code   (VITC) timecode encoded into the vertical interval of the video. It usually can be read out even when the VTR is still-framed or running at slower than play speed.

VHF   (very high frequency) the spectrum extending from 30–300 MHz

VHS   1/2-inch consumer-type video cassette format

Video conferencing using a computer or any network terminal (set-top box, kiosk) with a video camera attached to talk to and watch a person in a remote location. Video conferencing can employ specialist computer equipment and telecommunications infrastructures or can (with considerable loss of quality) be achieved over the Internet. Video conferencing is an established business tool, but video phones for the general public have not been successful. Flexible home ITV terminals and home PCs offer the user a cheap option to plug-in a video phone/conferencing camera and software, which may precipitate a mass market.

Video cassette recorder (VCR) classification of video recorders that include VHS, S-VHS, Beta, BetaCam, MII, as well as D-1, D-2, D-3, D-5, Digital BetaCam, Digital-S, DVCPro, BetaCam-SX, and DVCAM digital cassette recorders. The magnetic video tape is contained in a plastic case that is loaded into the VCR instead of on open tape reels (2-in, 1-in type C, etc).

Video on demand see VOD

Video dial tone see VDT

Video juke box see video server

Video server the combination of hardware and software that delivers multiple simultaneous channels of video

Virtual a seeming reality that lacks some elements of total reality

Virtual reality (VR) computer and video simulations that entail wearing headgear, electronic gloves, and possibly electronic body suits such that users are immersed in a cyberspace of simulated reality that gives the sensation of being in a three-dimensional world where objects can be moved about with hand movements and sensations of walking and touching are simulated using supercomputing power. Video games are the most successful VR applications because of the psychological engagement they stimulate with the environment. HDTV is another technology aimed at increasing the immersion effect by producing large screens that give the viewer the impression of being in the television world. One function of ITV could be to enable viewers to engage in virtual environments that are controlled centrally and populated by other users of the system. Networked video games have demonstrated this is an application with a considerable market.

Visits   a sequence of requests made by one user at one Web site. If a visitor does not request any new information for a period of time, known as the *time-out period,* then the next request by the visitor is considered a new visit.

VOD   (video on demand) an entertainment and information service that allows customers to order programs from a library of material at any time they desire. VOD systems demand that video servers must be capable of sending numerous streams of video simultaneously to different viewers. Pay-per-view VOD movie services are often predicted to be a major source of revenue for ITV services.

VPS   viewers per set

VPVH   viewer per viewing household

VRML   (Virtual Reality Modeling Language) Internet standard for 3D animations;   a computer language, similar to HTML, that allows Web site designers to create primitive forms of virtual reality on the Internet

VTR   video tape recorder

VTR format   a defined standard for tape recording information required in the TV production process, which outlines physical and electrical characteristics of the video tape system. Typical examples include VHS, DVCPro, Digital BetaCam, and D3.

Walled-garden Web browsing   a concept of providing ITV subscribers with access to a controlled set of valuable, aggregated Web sites, rather than providing them with the means to just browse the Web

WAN   (wide area network) a network of widely distributed (in different buildings or cities or countries) computers connected together through dedicated connections

WAP   (Wireless Application Protocol) this network-neutral protocol is used for sending data to and from WAP-capable devices, such as cellular phone handsets

Web common abbreviation for the World Wide Web on the Internet. The Web is a way of transferring files on the Internet using a hyperlink system to link computer files or pages with a simple user interface. The Web is based on a client-server system, where a piece of computer software called a browser can download and display text, graphics, and sound from any other computer on the network. The Web has become a huge collection of interlinked information and entertainment. Commercial Web sites act as marketing tools for companies. People "surf" the Web to search for information.

Web browser a piece of software that allows a computer user to look at information stored on the Web. See browser.

Webcasting use of Web to broadcast information. Unlike typical surfing, which relies on a pull method of transferring Web pages, webcasting uses a push combination of technologies to send information to users' computers. This is also referred to as *broadcasting* or *netcasting*. Users get steady updates of streams of information in requested categories. Users can subscribe to a channel, download software to a local computer, and then streams of automatic updates follow.

Web server a computer on which Web information is stored and served to other computers that request to view it

Web site a collection of information, using Web technology published by an individual or an organization and with a distinct address or URL

Web streaming live playback of media files on the Web

WebTV California company and its technology of the same name. WebTV is a way to provide Internet accesses, especially World Wide Web to the TV via a set-top box gateway over a telephone line. Company bought by Microsoft in April 1997, and technology licensed to several consumer electronic companies. Depends on revenues from subscription to a unique Internet service provider.

Whiteboarding a system that allows two or more people to share the same computer file on separate computers and to work on that file together at the same time

Wide screen   TV signal with a wider picture of 16:9 aspect ratio instead of conventional 4:3

Windows CE   a version of the Windows operating system designed for small devices such as personal digital assistants (or handheld PCs). The Windows CE graphical user interface (GUI) is very similar to Windows 95.

World Wide Web   see Web

Wrapper   a software module that surrounds other software, to fit the other software into the system. For example, a Java applet designed to display XML content embedded in traditional HTML documents

WWW   World Wide Web, allows users to have a simple interface to navigate and hyperlink to other information around the Internet. See Web.

About the Author

Dr. Hari Om Srivastava received his M.Sc. in electronics from Gorakhpur University in 1967, a Ph.D. in chemistry from Poorvanchal University in 1996, and another Ph.D. in information systems from BITS, Pilani, India in 1999. He joined the Indian Broadcasting Service in 1972 and has been responsible for managing and expanding India's broadcast network. Dr. Srivastava has received training in information technology in the United Kingdom, the United States, Japan, and Norway. He worked as a Commonwealth expert in 1991 and an ITU expert in 1992. Dr. Srivastava has written dozens of articles and research papers for national and international journals, and in 2000 he published *Broadcast Technology: A Review.* He has received two national and six international awards and appeared in the *International Who's Who* in 1998 and the *World's Who's Who* in 2001. Dr. Srivastava is currently chief engineer and head of air resources in the Directorate General of All India Radio. His areas of interest include information systems, digital storage, multimedia broadcasting, and broadcast applications for social issues.

Index